T0052615

The Eye: A Very Short Introduction

VERY SHORT INTRODUCTIONS are for anyone wanting a stimulating and accessible way in to a new subject. They are written by experts, and have been translated into more than 40 different languages.

The Series began in 1995, and now covers a wide variety of topics in every discipline. The VSI library now contains over 350 volumes—a Very Short Introduction to everything from Psychology and Philosophy of Science to American History and Relativity—and continues to grow in every subject area.

Very Short Introductions available now:

WITTGENSTEIN A. C. Grayling
WORK Stephen Fineman
WORLD MUSIC Philip Bohlman

THE WORLD TRADE
    ORGANIZATION Amrita Narlikar
WRITING AND SCRIPT
    Andrew Robinson

Available soon:

NUTRITION David A. Bender
CORAL REEFS Charles Sheppard
COMPLEXITY John Holland

HORMONES Martin Luck
ALEXANDER THE GREAT
    Hugh Bowden

For more information visit our website
www.oup.com/vsi/

Michael F. Land

THE EYE

A Very Short Introduction

OXFORD
UNIVERSITY PRESS

Great Clarendon Street, Oxford, OX2 6DP,
United Kingdom

Oxford University Press is a department of the University of Oxford.
It furthers the University's objective of excellence in research, scholarship,
and education by publishing worldwide. Oxford is a registered trade mark of
Oxford University Press in the UK and in certain other countries

© Michael F. Land 2014

The moral rights of the author have been asserted

First edition published in 2014

All rights reserved. No part of this publication may be reproduced, stored in
a retrieval system, or transmitted, in any form or by any means, without the
prior permission in writing of Oxford University Press, or as expressly permitted
by law, by licence or under terms agreed with the appropriate reprographics
rights organization. Enquiries concerning reproduction outside the scope of the
above should be sent to the Rights Department, Oxford University Press, at the
address above

You must not circulate this work in any other form
and you must impose this same condition on any acquirer

Published in the United States of America by Oxford University Press
198 Madison Avenue, New York, NY 10016, United States of America

British Library Cataloguing in Publication Data
Data available

Library of Congress Control Number: 2014930361

ISBN 978-0-19-968030-6

Printed and bound by
CPI Group (UK) Ltd, Croydon, CR0 4YY

Links to third party websites are provided by Oxford in good faith and
for information only. Oxford disclaims any responsibility for the materials
contained in any third party website referenced in this work.

Contents

Acknowledgements

I would like to thank the editorial team at Oxford University Press, particularly Latha Menon, for encouragement and advice on the text. Also my daughter Kate Land, who read all the text as an educated reader without specialist knowledge, and made numerous helpful suggestions; and to an optometrist friend, Colin Davidson, who helped with the final chapter on impaired vision.

List of illustrations

Springer and Plenum Press, New York, *Eye Movements and Vision*, 1967, pp. 171–77. *Eye Movements during Perception of Complex Objects*, A. Yarbus, with kind permission from Springer Science + Business Media B.V.
After K. Brodmann (1914) Physiologie der Gehirne. In *Die Allgemeine Chirurgie der irnkrankheiten, Neue Deutsche Chirurgie*, Vol.11: *Enke*. From J. P. Frisby (1979) *Seeing*, p. 79. Oxford University Press
Superstock/Glowimages.com

The Eye

Chapter 1
The first eyes

Origins

Of all the senses, vision is the most versatile. It allows animals to navigate through the environment, seek out food, and detect and avoid predators. Good eyesight makes it possible to recognize other individuals and to communicate with them by gesture and expression. Hearing, smell, and touch can each fulfil some of these functions: sound is useful in communication, the chemical senses can identify food, and a rat in the dark can navigate by touch. But for many animals vision predominates, and its loss is more devastating than the loss of any other sense. Eyes come in many varieties, from the simple pit eyes of flatworms to the sophisticated compound eyes of arthropods and the single chambered eyes of vertebrates and cephalopod molluscs such as *Octopus*. These are discussed later in the chapter, and photographs of some of the more remarkable eyes are shown in Figure 7 towards the end of the chapter.

It seems that the evolution of eyes started very slowly in the Precambrian period and then took off during the Cambrian, between 541 and 485 million years ago. Since that time many refinements have occurred, but all the basic designs were in place by the end of the Cambrian (Figure 1). The earliest eyes for which we have a fossil record were not single-chambered eyes like ours,

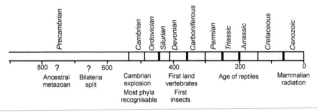

1. Geological time-line. Age scale in millions of years before present. Five major extinctions, in each of which more than 70% of the world's fauna disappeared, are shown as thicker bars

but were the compound eyes of trilobites. These animals had an external skeleton and scavenged the sea floor about 520 million years ago, until the great extinction at the end of the Permian, 270 million years later. The reason we know more about these eyes than any others is that their lenses were made of the mineral calcite, so they effectively came pre-fossilized. Animals with our kind of eye—the first fish-like creatures and the cephalopod molluscs—also evolved in the Cambrian, but whatever eyes they had have not fossilized. Although the Cambrian was a period when animals became more mobile and vision became a major sense, there are good reasons for believing that animals had some visual capabilities earlier than this. Trilobite eyes are already quite sophisticated, and must have had simpler antecedents. Stronger evidence comes from the photoreceptors themselves—the cells that respond to light and make vision possible. These are present in jellyfish, whose origins go further back into the Precambrian. Sadly, providing dates for evolutionary events before about 555 million years ago is fraught with problems, and estimates for the timing of the beginnings of multicellular animal life vary by at least a hundred million years.

What we can say, however, is that all animals, from jellyfish to man, share a particular molecule—rhodopsin—which is responsible for starting the process of converting light into the electrical signals that the nervous system can make use of. Plants, fungi, and even bacteria also have light receptive molecules, but none resemble animal rhodopsin, and we can assume from this rhodopsin evolved at much the same time as multicellular animal life. Rhodopsin is a two-part molecule. It consists of a protein (opsin) and, held within it, a smaller molecule—the 'chromophore'—related to vitamin A (Figure 2). This chromophore has a long chain of double bonds that

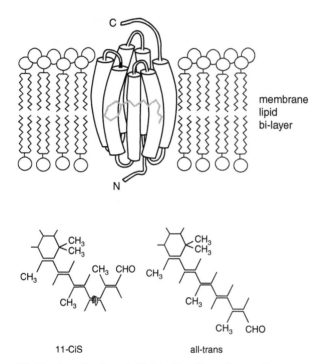

11-CiS   all-trans

2. **Rhodopsin molecule embedded in the cell membrane of a photoreceptor. The membrane is a double layer of lipid molecules. Rhodopsin consists of an opsin protein, with 7 helices of amino acids that cross the membrane, and inside it (in grey) sits the chromophore molecule—usually a relative of vitamin A. This is shown below in its two forms: the 11-cis form before it is stimulated, and the all-trans form it converts to when a photon of light is absorbed. This conversion starts the visual process**

3

are tuned to respond to light energy. When it absorbs the energy of a photon of light its structure changes, and this change initiates a series of biochemical reactions (known as the transduction cascade) that ultimately results in an electrical change in the photoreceptor cell, which is then transmitted to the nervous system. In all animals, this is the first step in vision.

A quick summary of early animal evolution

Before discussing the different kinds of photoreceptors and eyes it will be useful to have an outline of where eyes fit into the basic narrative of animal evolution (Figure 3). Of the earliest animal groups (phyla) the sponges (Porifera) do not have photoreceptors or anything remotely eye-like, so photoreception was not universal among the first animals. However, many animals of another early

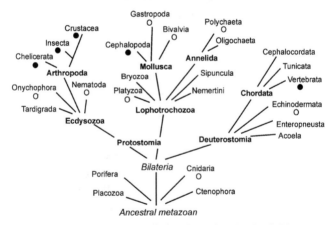

3. **Evolutionary tree of the animal kingdom, showing the division of the Bilateria into three main superphyla, the Ecdysozoa, Lophotrochozoa, and Deuterostomia. Groups with eyes of some kind are shown with circles (○). Groups with eyes that resolve well have filled circles (●): these evolved only in the arthropods, the cephalopod molluscs, and the vertebrates. Latin names are used here, but convert easily to English equivalents by replacing the last 'a' with an English plural**

phylum, the Cnidaria, which contains the sea anemones, corals, and jellyfish, do indeed have photoreceptors. In one class of cnidarians, the cubozoans that include the notorious Australian 'stingers', there are even eyes that really look like eyes, so the capacity to produce eyes was present from a very early stage of metazoan evolution. The Cnidaria, which are radially symmetrical animals, split off from the early metazoan line before the formation of the main animal phyla, collectively known as the Bilateria from their left–right symmetry. Most of these bilaterian animals moved forwards, and so had a head end where most receptors, such as eyes, were located. At some time in the late Precambrian the bilaterians split into three major groups, the Ecdysozoa, the Lophotrochozoa, and the Deuterostomia (Figure 3). Excellent eyes evolved in at least one branch of each group. Before the advent of molecular ways of tracing animal relationships only two major bilaterian groupings were recognized: the protostomes and the deuterostomes. The division was based on their different development: the names mean first mouth and second mouth, from the way the mouth and anus develop in the early embryo. The second of these groups—the Deuterostomia—is the group to which we humans belong, and contains the starfish, the sea squirts, and the chordates, from which the vertebrates arose. In 1997, the Protostomia were divided into their two present groups, the Ecdysozoa and the Lophotrochozoa, on the basis of new molecular evidence. Basically the Ecdysozoa are animals that moult repeatedly as they grow. They include the jointed-limbed animals (arthropods), and these comprise the Chelicerata (horseshoe crabs, scorpions, and spiders), the Crustacea (shrimps, lobsters, crabs, and many smaller classes) and the Insecta (beetles, flies, etc.). Most of the members of these groups have compound eyes and good—sometimes excellent—vision. The Ecdysozoa also include millipedes and centipedes, and perhaps surprisingly the nematode worms, which also moult but are mostly eyeless. The Lophotrochozoa (an unconvincing hybrid word meaning 'crest/wheel animals') include most of the rest: the flatworms, the molluscs, the annelid worms, and several smaller phyla. Of these, the cephalopod molluscs (octopus, squid, and cuttlefish) stand out as having eyes that are large, resolve well, and are similar in their capabilities to those of fish.

The first photoreceptors

Not surprisingly, photoreceptors evolved before eyes. There are basically two types of specialized photoreceptor cell in animals, known as 'rhabdomeric' and 'ciliary', and for many years it was thought that this distinction mapped nicely onto the protostome/deuterostome divisions of the Bilateria. To catch a reasonable proportion of the light that falls on them, photoreceptors need a high density of rhodopsin. Rhodopsin molecules are embedded in the membranes of the receptor cells (Figure 2) so increasing the rhodopsin density means having a greatly expanded membrane. Ciliary receptors are organized around cilia, hair-like structures found in a wide variety of cells. Typically the membrane is organized into stacks of lamellae or discs, as in the rods and cones of the human retina (Figure 4). A human rod has a total of

4. Rhabdomeric and ciliary photoreceptors, based on a receptor from the fruit-fly *Drosophila*, and a human rod. N is the nucleus. At the base of each receptor is the synaptic junction that connects with the nervous system

150 million rhodopsin molecules packed into about 1,000 discs. In rhabdomeric receptors there is no cilium, and the membrane is expanded into finger-like extensions known as microvilli, closely packed together. In arthropods these almost crystalline structures are known as rhabdoms (Greek for 'rod'). The two receptor types differ not just in structure but in the molecular make-up of the rhodopsin molecules themselves, as well as in the biochemistry of the transduction cascades. So it seems that there are two quite distinct photoreceptor families which must have diverged at an early stage in animal evolution, or else evolved from quite separate receptor structures.

In recent years it has become clear that the protostome-rhabdomeric and deuterostome-ciliary division is not as rigid as first thought. There are some rhabdomeric-type receptors in the human retina—not in the rods and cones but among the ganglion cells that transmit impulses to the brain. In the protostomes, ciliary receptors are found in the brains of worms, and the eyes of scallops have both types of receptor. Thus protostomes and deuterostomes contain receptors from both families, in one capacity or another, and this must mean that both types were present in the early bilaterians, in other words before the protostome–deuterostome split, and well before the evolution of most of the structures we would regard as eyes. The photoreceptors of cnidarians, which preceded the bilaterians, are neither quite one type nor the other. They are possibly more like ciliary than rhabdomeric photoreceptors, but the jury remains out on this.

The emergence of eyes

What constitutes an eye, as opposed to a simple photoreceptor, has never been entirely clear. Some authors limit the term 'eye' to the highly complex structures of vertebrates, cephalopods, and arthropods, while others apply the same word to any photoreceptor beyond the one-cell stage. In this book, I will use the following definition.

An eye must contain at least two photoreceptors with different fields of view, so that comparisons can be made simultaneously between light from different parts of the environment.

Eyes do not have to be good to be useful. A simple photoreceptor, with no screening pigment or optics, can be used to tell day from night, determine depth in the sea or closeness to the surface for a burrowing animal, or to retreat from predators that cast a shadow. By slowing down or speeding up locomotion it can be used to allow an animal to remain in, or escape from, a particular light environment (kinesis). By adding screening pigment on one side, such a photoreceptor can be used to tell the rough direction of light, and if the receptor is waved from side to side, sampling different directions, it can be used to guide an animal towards or away from light (klinotaxis). This is what fly larvae do, and is rather like the way we would use our nose to track down the source of a bad smell. With two such detectors, one on each side of the head, an animal can estimate light direction directly by comparing the two outputs, and steer accordingly (tropotaxis). So far, none of this requires an eye, as defined above.

At some early stage in the evolution of bilaterian animals two basic eye designs emerged. These are single-chambered eyes, of which our eyes are descendants, and compound eyes that gave rise to the wide variety of faceted eyes of crustaceans and insects. Single-chambered eyes (often misleadingly referred to as simple eyes) have their receptors in a single concavity, or pit (Figure 5); compound eyes have them in outward-pointing pigment tubes distributed over a convex surface (Figure 6). As there seem to be very few intermediates I will deal with the two eye types separately.

Single-chambered eyes

The control of all the activities mentioned in the last section can be improved by having several receptors in each pigment cup,

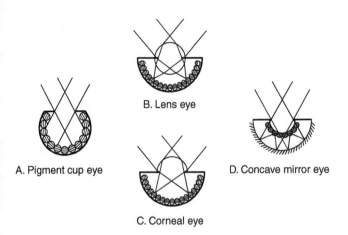

5. Diagrams of the four main types of single-chambered eye. These show the paths of rays coming from two distant points. The photoreceptors are shown cross hatched and mirrors in diagonal hatching

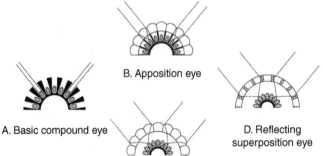

6. The main types of compound eye. Receptors are cross-hatched and mirrors have diagonal hatching

making a basic eye (Figure 5a). Even without optics, a cup with an aperture that partially restricts the entry of light will ensure that single receptors view different regions of the surroundings. This is the first stage in the development of an imaging system, and although the field of view of single receptors might be 30° or more, this represents a considerable advance on the 180° field of a single pigment-backed receptor. Such eyes are found in many platyzoans (flatworms), molluscs, and annelid worms, and structures like these must have been the starting point from which more advanced eyes, such as those of vertebrates, evolved. These eyes can be used to avoid obstacles, to detect self-motion, and to allow orientation to coarse landmarks and major celestial objects such as the sun or moon. For many of the small animals that live between the tides, these are the tasks that matter, and this kind of simple eye is all they need.

To go beyond this stage, animals had several options. They could close down the aperture of the cup to produce a pinhole eye; they could evolve a lens to concentrate light onto smaller regions of the retina; or they could use a concave mirror to produce an image. Alternatively, they could go down the compound eye route, as discussed later. Pinhole eyes are rare, for good reason. Closing the aperture down improves resolution, but cuts down the light entering the eye to barely useable levels. There are a few small pinhole eyes in some gastropod molluscs, such as the abalone, and in giant clams, where their function is to spot fish approaching, as these might nibble the mantle tissue that protrudes from the shells. More impressive are the eyes of the cephalopod *Nautilus* (Figure 7a), an ancient relative of octopus and squid. These eyes are large (1 cm), have a pinhole with a variable aperture, and muscles to keep the eye upright. How *Nautilus* failed to evolve a lens, when all its relatives succeeded, remains a mystery.

Having a lens solves the problem of this trade-off between resolution and the amount of light reaching the retina. Initially a lens would not have to form a proper image to improve things.

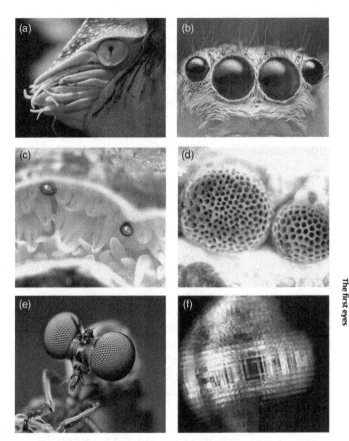

7. A selection of animal eyes: (a) The large pinhole eye of the cephalopod mollusc *Nautilus*. (b) Corneal eyes of a jumping spider *Platycryptus*. (c) Concave mirror eyes of the scallop *Pecten*. (d) Primitive compound eye of the ark clam *Barbatia*. (e) Compound eye of a male robberfly *Holocephala*. (f) Mirror compound eye of the shrimp *Palaemonetes*

Filling the eye with some substance with a refractive index higher than sea-water, perhaps dense mucus, would narrow the cones of light seen by each receptor. Over time, more highly refractile substances such as crystallins could be substituted and the 'lens'

11

condensed into a sphere, so that eventually a properly resolved image would be produced (Figure 5b). In the gastropod molluscs one can trace a series of eyes in existing forms that shows this sequence, from limpets with simple pit eyes to periwinkles with eyes that have excellent lenses. The subtleties of how animals make lenses free from optical defects will be discussed in the next chapter.

In the final stage in the evolution of single-chambered eyes, resolution improved to the extent that single receptors were viewing angles smaller than one degree (approximately the angular size of one's little fingernail at arm's length). With such eyes predation and predator avoidance became possible, as did the recognition of possible mates, navigation at speed, and the full range of visually guided activities that we associate with vertebrates. For single-chambered eyes this happened in only two groups, the cephalopod molluscs and the chordates (Figure 3). True cephalopod molluscs from which modern groups—octopus, squid, and cuttlefish—derive certainly evolved in the late Cambrian. A small chordate, *Pikaia*, resembling a fish larva, was also present in the Cambrian fauna. Whether it had eyes isn't clear, but a Cambrian relative, *Haikouichthys*, almost certainly did. About 30 million years later, in the Ordovician, another group of chordates, the conodonts, had larger eyes. Conodonts are classed as vertebrates, but their relations to other fish groups are unclear: they died out 200 million years ago, at the end of the Triassic (Figure 1). However another early fish group from the Ordovician, the agnathans—jawless fish—do have living relatives, the hagfishes and the lampreys. Hagfish are almost eyeless, but the lampreys do have good eyes. Disappointingly, the lamprey eye is so similar to the eyes of all other fish that it tells us little about the early evolution of vertebrate eyes, except that the present design is more than 400 million years old. So although we have quite a good account of the presumed evolutionary sequence of eyes in molluscs, there is no equivalent series for the eyes of vertebrates.

Vertebrates evolved onto land during the late Devonian, about 370 million years ago (Figure 1). When they did, the cornea, which until then had simply been a protective cover for the eye, became a refractive structure because it now had air on one side and fluid on the other (Figure 5c). In land vertebrates the cornea largely took over the image-forming role from the lens, whose function then became more one of adjusting focus (accommodation). Otherwise the eye structure remained much the same as it had been in marine vertebrates. The only other large animal group with eyes of the corneal type are spiders. The best of these, the jumping spiders, have eyes that resolve better than any other invertebrate (Figure 7b), apart from the cephalopod molluscs.

A very small number of animals have single-chambered eyes based not on a lens but a concave mirror (Figure 5d). The best-known of these are the eyes that surround the mantle of scallops (Figure 7c), and act as movement detectors that allow the animal to close its shell before its tentacles are bitten off—a similar 'burglar alarm' function to the pinhole eyes of giant clams. Here the retina is in the middle of the eye, and light passes through it before coming back, focused, from the hemispherical mirror at the back. A few crustaceans use the same principle, but it is not common, probably because the unfocused light entering the eye reduces the contrast of the focused image.

Compound eyes

In early compound eyes each receptor would have had its own pigmented tube, excluding light from all but one direction (Figure 6a). Such eyes still exist, in ark clams (Figure 7d), where, like the eyes of scallops and giant clams, they serve to detect the motion of predators. Adding a lens to the mouth of each tube improves performance by getting more light to the receptors, and better defining each receptor's field of view (Figure 6b). This type of eye, known as 'apposition' because the overall (erect) image is made up of the apposed contributions of all the individual receptors, was present in the first trilobites, and is still the most common type of eye in diurnal insects (Figure 7e), many crustaceans, and some centipedes. It was probably the ancestral eye of chelicerates—the horseshoe crab *Limulus* still has such eyes—before becoming simplified into a set of corneal single-chambered eyes in spiders.

In nocturnal insects, and some marine crustaceans such as krill, a new type of eye evolved from the apposition eye. These 'superposition' eyes have a much larger pupil, with much of the eye surface contributing a very bright erect image which is formed on a deep-lying layer of receptors (Figure 6c). The optics are not simple. In 1891 Sigmund Exner, the father of compound eye research, showed that to produce a superposition image each optical element has to behave not as a single lens, but as a two-lens telescope that inverts the direction of the light. He went on to show that for this to happen light has to be bent within each element by gradients of refractive index. So tricky was this idea that Exner was not finally proved right until 1973. Shortly after this, in 1975, Klaus Vogt showed that crayfish had a different sort of superposition eye based on radially arranged mirrors (Figure 6d). In these eyes light is reflected from one, or usually two, sides of square box-like structures lined with mirrors, which redirect light to a common focus much as in their refracting counterparts. This type of eye is confined to the more edible decapod crustaceans—the shrimp (Figure 7f), crayfish, and lobsters—but interestingly not the crabs, which have stayed diurnal and retained apposition eyes.

Compound eyes do not resolve particularly well, because the small size of the lenses means that diffraction limits the minimum resolvable angle to between 0.5 and 5° (see Chapter 2). This, however, does not prevent insects from having a full range of visual behaviours, from landmark navigation to interception on the wing. Insects also have varieties of vision that we lack, such as the ability to see ultra-violet colours, and to navigate using the pattern of polarization in skylight.

The time course of eye evolution

Darwin was famously worried that the human eye was so complex and well engineered that natural selection might not have been up to the task of producing it. Although, a few paragraphs later in the *Origin of Species* he dismissed this troubling thought, there remained, nonetheless, a question of how long it would take such complex structures to evolve. Was there, for example, enough time in the few million years of the Cambrian explosion for complex eyes to come into being from structures like those of Figures 5a and 6a? This question was addressed directly by Dan-Eric Nilsson and Susanne Pelger in 1994. They produced a genetically realistic model of the progression from a light-sensitive patch of skin to a typical fish or cephalopod eye, subject only to selection which favoured increased spatial resolution. Their model assumed typical values of variation within a population, and an inter-generation time of a year, which is reasonable for a small animal. Their conclusion was that the complete series of transformations could be achieved in about half a million generations, corresponding roughly to half a million years, which is quite short enough to have allowed eyes to change at a rate consistent with their rapid evolution in the Cambrian. It is a conclusion that would certainly have pleased Darwin.

Chapter 2
Making better eyes

Producing high quality images

Human eyes can resolve one minute of arc. This is an angle corresponding to a grating of 1cm lines viewed at a distance of 34 metres. Comparable resolution can be found in many other vertebrates and cephalopod molluscs. This requires quite sophisticated optics. In optical technology lenses are made of multiple components of different glasses, and their complex structure is designed to combat the imperfections in the image that result from refraction at simple curved surfaces. In biological optical systems the same results have to be achieved in a different way, using the rather improbable materials of protein and water. In addition to producing a well resolved image on the retina, advanced eyes must also be capable of providing the brain with useable information over a huge range of light levels. This chapter explores some of the difficulties associated with making versatile high-resolution eyes, and how these problems have been overcome.

Eyes evolved in the sea and, with no corneal refraction to help them, they needed lenses that were as powerful as possible so that light could be brought to a focus a reasonably short distance behind the lens. In practice this meant having a spherical lens, as a sphere has the highest possible surface curvature and hence optical power. Although animals do not make glass (trilobites with calcite lenses came closest to this), dry protein is a highly refractile material, and is commonly available. It has a refractive index of about 1.53, which compares reasonably with glass (1.52–1.63). However, a spherical lens made of glass, or dry protein, produces a very poor image. The reason for this is that rays going through the lens close to the central axis come to a focus at one point (the one predicted by the simple lens formula), but rays further from the axis are focused progressively further in front of this 'ideal' point, and so the image becomes a blur. This is known as spherical aberration, and is so serious that such a lens would be quite unusable (Figure 8a). If the refractive index decreases from centre to periphery, essentially by diluting the protein with water, then refraction occurs within the lens and the more peripheral parts of the lens bend light less, so that all rays come to a common focal point (Figure 8b). For this to happen, the gradient of refractive index must be correct. Working out the exact form of this gradient proved difficult, and a solution was not finally achieved until 1944. Such a gradient has a further benefit in that it reduces the focal length of the lens as a whole.

During the 1880s Ludwig Matthiessen measured the focal lengths of the lenses of many fish and cephalopod molluscs, and he found

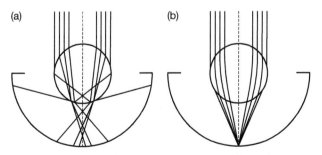

8. (a) Lens with a homogeneous refractive index, showing the effects of spherical aberration. (b) Lens with a graded refractive index in which all rays come to a common focus

that in all these the ratio of lens focal length to radius was close to 2.5:1; this has become known as Matthiessen's ratio. He knew this meant that these were gradient lenses, because a lens made of homogeneous protein would have a focal length of about 4 radii. Matthiessen himself believed the gradient to be parabolic, which, although not correct, was not a bad approximation. It now seems that virtually all the spherical lenses found in marine animals, including some unlikely ones in annelid worms and copepod crustaceans, conform to Matthiessen's ratio, and this means that all these animals have hit on the same trick of building a gradient into the lens. Since any small change towards a gradient will improve the image, it is perhaps not surprising that once animals had embarked on this evolutionary path they followed it to its logical conclusion.

A second problem with biological lenses is chromatic aberration. All refractive structures bend blue light more strongly than red light, so that the image for blue light is closer to the lens than the red image. The separation of these images is quite large, about 10 per cent of the average focal length, which in turn means that the 'best' image, somewhere between the blue and red images, is a blur. For a colour-blind animal, such as an octopus, or a fish living in the deep sea where only blue light penetrates, this is not a problem: only one part of the spectrum is involved, and the retina can be positioned appropriately. Shallow-water fish, however, have excellent colour vision, with several visual pigments tuned to different wavelengths across the spectrum. There is apparently no single plane where receptors with different pigments would all receive a sharp image. Putting different receptors at different distances from the lens would work, but in a fish retina the receptors do in fact all lie in a single layer. A partial solution to this conundrum was proposed by Ronald Kröger and his colleagues in 1999. It appears that in real lenses the gradient of refractive index, which, as we have seen, corrects for spherical aberration, is not quite the same as the 'ideal' gradient. It is tweaked so that different zones of the lens produce images at slightly different

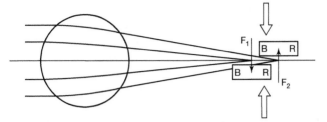

9. Lens with two foci, at F_1 and F_2 spaced so that the red image from F_1 coincides with the blue image from F_2. In this plane (arrows) there is a sharp image for both wavelengths

distances (Figure 9). Each of these images suffers from chromatic aberration, but if they are spaced correctly the 'blue' image from one zone can be in the same plane as the 'red' image from another zone. In other words, there will be a plane where there is a sharp image for all relevant wavelengths. This is not a perfect solution because the in-focus blue image from one lens zone will always be accompanied by an out-of-focus blue image from another zone, and similarly for other wavelengths. This effectively lowers the contrast of the image the receptors receive, *but it is still a sharp image*, and evidently the reduced contrast is a price worth paying. In optical technology the chromatic aberration problem is overcome by using doublet lenses made of glasses with different dispersions (i.e. where the difference between the refractive indices for red and blue light are not the same) so that variations of focal length with wavelength are cancelled, but it seems that biological materials offer no such possibility.

Spherical and chromatic aberrations affect the corneal eyes of land animals just as they do the lens eyes of fishes. In humans the cornea would produce spherical aberration if it were a spherical surface: but, at least in humans, it is not. Our cornea has a profile in which the curvature in the centre is stronger than it is in the outer regions, so that, just as with the corrected spherical lens, rays further from the central axis are bent relatively less, and in this way all rays are brought to a common focus. The lens itself has a graded index, inherited from the lenses of our fishy ancestors. It seems, however, that chromatic aberration in the human eye is too small to be worth correcting. There are two reasons for this. First, our pupil in daylight is small, so that the cone of light reaching the receptors is narrow and the shift in image distance with wavelength produces much less blur than with the relatively huge apertures of fish eyes. Second, we resolve detail best in a relatively narrow range of wavelengths in the red-green part of the spectrum, and very poorly in the blue, so we do not need an image that is corrected across the spectrum.

Diffraction: the fundamental limit

The aberrations just described can be 'fixed', to a greater or lesser extent, by varying gradients and curvatures. There is, however, one image defect that cannot be cured this way. When light from a distant point is focused by a perfectly corrected optical system it does not form a point image but a 'blur circle', in which the distribution of light in the image is as shown in Figure 10. This occurs because light waves from different parts of the aperture interact with each other to produce an interference pattern, known as an Airy disc (after its discoverer in 1835, William Biddell Airy). The width of this disc, at half its maximum intensity, is given simply by λ/D: the wavelength of light divided by the aperture diameter. This gives the disc width as an angle in space outside the eye, but it can be easily converted to a distance on the retina by multiplying by the eye's focal length. The size of this pattern on the retina of a human eye, with a 2.5 mm pupil and light with a wavelength of 0.5 μm (blue-green), works out to be 3.34 μm, slightly larger than the width of the receptors (the cones) in the centre of the retina. Notice that, since the width of the Airy disc is *inversely* proportional to D, this means that the larger the aperture of the eye the better the image should be. The problem here is that the *other* aberrations—spherical and chromatic aberration, and more exotic ones like astigmatism and coma—all get worse as the aperture increases. Thus humans have their best

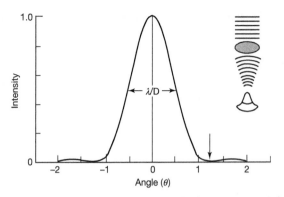

10. **Distribution of intensity across the Airy diffraction pattern. The width of this pattern, at half its maximum intensity, depends on the wavelength of light (λ) divided by the aperture diameter (D). The inset shows how light waves from a distant point are bent by the lens and converge on the image where they interfere to produce the Airy disc pattern**

resolution when the aperture—the pupil—has a diameter of about 3 mm. In dim conditions, when the pupil does open beyond this to a maximum of about 8 mm, the resulting loss of resolution is considerable, but it barely matters, because the lack of light limits resolution for other reasons.

The diffraction limit is more of a problem for insect eyes. Because of the way they are constructed, with one lens for each directional element in an apposition eye (Figure 6b) the lenses are necessarily tiny. In a bee they are about 25 μm in diameter. This is 100 times smaller than the daylight aperture of our eye, and the image is accordingly 100 times less acute. A bee's resolution is about 1°, an angle corresponding to a fingernail at arm's length, compared to less than 1 arc-minute in ourselves. Remarkably, there is not much that this lack of detailed vision stops bees from doing: a coloured flower will still be visible at a distance of several metres, and even blurry landmarks can act as reliable guides. Few insects have lenses much bigger than those of bees. Some, like dragonflies, manage to squeeze in a region of enhanced resolution, but to increase

21

resolution all round leads to a ludicrously large eye. This is because both the size *and number* of facets must increase, so eye size increases as the *square* of resolution. (In single-chambered eyes like ours size is simply proportional to resolution.) It is one of the mysteries of evolution that adult insects didn't opt at some stage for the single lens type of eye, particularly as these are actually present in many insect larvae. As Dan-Eric Nilsson put it:

> It is only a small exaggeration to say that evolution seems to be fighting a desperate battle to improve a basically disastrous design.

The size of receptors

As in photography, the detail in an image can only be recorded fully if the optical quality of the lens and the fineness of the mosaic of receiving elements are roughly matched. If these elements are too wide detail will be lost, and equally there is no point in having elements smaller than the finest detail in the image. In the eye it takes two receptors to resolve a black and white line pair in the image of a grating, and the finest line pair a diffraction limited eye can resolve is equal to the width of the Airy disc (λ/D; Figure 10). In the human eye in daylight this is an angle of 0.69 arc minutes, or, as mentioned earlier, a distance of 3.34 μm on the retina, so one might expect the receptor spacing to be half this: 1.67 μm. In fact the cone spacing in the centre of the fovea is slightly greater than this: 2.5 μm. One reason for the apparent mismatch may simply be that receptors with these dimensions are already as narrow as it is possible for them to be. Photoreceptors act as light-guides, so that once light gets inside it remains trapped by reflection from the outer membrane. However, as their diameter decreases towards the wavelength of light (0.5 μm) light-guides become 'leaky', and some light starts to travel down the outside of the guide, where it becomes available for other receptors to capture. This cross-talk blurs the quality of the received image, and means that it is pointless to have receptors much narrower than about 2 μm. In fact very few receptors do have a diameter less than 2.5 μm, in vertebrates or in other phyla, and this width limit imposes a real restriction on the way eyes can be constructed: an eye cannot achieve higher resolution by having larger numbers of narrower receptors, so the only way to do it is to become larger. In dim light these limits do not really apply. Here the image is degraded not by the optics or the receptor mosaic, but by the lack of light. Under these circumstances it makes sense for the receptors to widen, or for many to pool their outputs to collect more light and so provide a better signal.

Dim light

We can see something over an intensity range of roughly 10^{10}, from bright sunlight to overcast starlight. Moonlight is about a million times dimmer than sunlight, and colour is no longer visible, as the cones in the retina no longer provide a useable signal. The rods, however, carry on for another four decades of decreasing intensity until they also have too little light to function. The problem of seeing in dim conditions is that light is quantal and arrives as indivisible photons that are either absorbed by the rhodopsin molecules, or not. Rods absorb and respond to single photons, but in dim light the number of photons available is very low: at the threshold of vision single rods absorb photons at a rate of less than one per minute. The energy involved in photon absorption by rhodopsin molecules is tiny. M. H. Pirenne, in his book *Vision and the Eye*, illustrates this dramatically by comparing light and mechanical energy:

> The mechanical energy of a pea falling from a height of one inch, would, if transformed into luminous energy, be sufficient to give a faint impression of light to every man that ever lived.

Rod receptors are thus as sensitive as is physically possible. However, the brain does not accept a single rod response as proof of the absorption of a photon, because there are other reasons unrelated to light which might cause a rod to respond.

The safety factor involved is a requirement that about six rods respond together, before a positive identification of a dim flash can be made.

The situation at light levels above threshold is well illustrated by Figure 11, taken from Pirenne's book. It represents a patch of retina with 400 rods, first at the threshold of vision (I: 6 absorptions) and then at three light levels each increasing by a factor of ten. The task is to determine when it is possible to

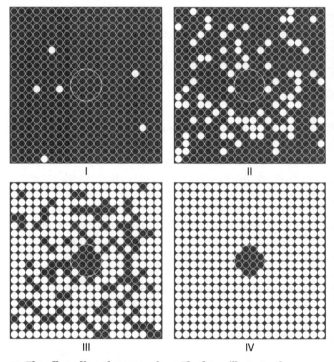

11. **The effect of low photon numbers. The figure illustrates the capture of photons by a field of 400 rods at the threshold of vision (I), and at light levels which each increase by a factor of ten. At what light level can the central dark disc be made out with certainty?**

make out the presence of a black disc. In II there is not enough evidence. In III a gambler might chance it, but it is only in IV, where the intensity is a thousand times the threshold, that one can be certain. At low light levels the absorption of photons is random, and therefore the problem of what can be seen becomes a statistical question. As with other statistical problems the quality of the information that can be derived from the distribution, of photons in this case, is a function of the number available. Fortunately the statistics of photon numbers are well understood, and it is possible to calculate how the representation of a scene will improve as numbers increase. In Figure 11 the pattern is black and white, but in a normal scene it will be composed of shades of grey, and one can calculate how well the different grey levels can be resolved, and by how much image detail will be degraded, as a function of light level. It turns out that all aspects of vision continue to improve until the rate at which photons reach receptors is well into the millions per second: light levels corresponding to a well-lit room. We are thus somewhat photon starved at all levels except the very brightest. There are good reasons for insisting on adequate lighting for places of work and study.

In terms of their eye structure, humans are basically diurnal animals. We have an optical system with a small aperture which provides a very well resolved image in daylight. Only birds of prey have significantly better resolution than we do. Many animals are crepuscular or even nocturnal in their habits, and fish in the deep sea capture prey at light levels that are well below the human threshold of vision. How can eyes be adapted to function at light levels where we would barely be able to get around? Even in our eyes, the photochemistry of vision is tuned to work at the physical limits of sensitivity, so not much more can be done there. The crucial adaptations are twofold: to get as much light onto the image as possible, and to sum the signals from many receptors over both space and time to provide a better signal, while retaining adequate resolution. Improving image brightness is largely a matter of increasing the effective aperture of the eye: specifically the ratio of aperture to focal length (D/f). In cats and dogs the lens is relatively larger than in humans, and in nocturnal rodents it almost fills the eye. However, the lens cannot be larger than the eye itself, so the gain in image brightness resulting from enlarging the lens is limited to a factor of not more than about ten. A way round this is to make the whole eye bigger, and certainly some of the largest eyes in vertebrates are found in animals that are active or at least vigilant at night. Horse eyes, for example, are huge even though their resolution is not special—about three times worse than in man; their eyes are larger than those of elephants. The largest eyes of any animal are the dinner-plate-sized eyes of giant squid, which hunt prey and detect predators at a depth below 500 metres in the ocean, where light is almost limited to the bioluminescence of other animals. With a large eye, although the image brightness is no greater because the ratio D/f does not increase with eye size, the size of the pool of receptors corresponding to a pixel in the image can be made bigger, without losing resolution. Vertebrates generally have the capacity to vary the size of this pool depending on the light level, by changing connexions in the retina. The other way of increasing the size, and hence reliability, of the visual signal is to sum the receptors' responses over longer periods of time. Vision thus becomes slower in the dark, and this increased summation time shows up, for example, in the long traces of sparklers waved in the dark.

A final comment

In a discourse about the human eye in 1868, the German polymath Hermann von Helmholtz wrote:

> For the eye has every possible defect that can be found in an optical instrument, and even some which are peculiar to itself; but they are all so counteracted, and the inexactness of the image which results

from their presence very little exceeds, under ordinary conditions of illumination, the limits which are set to the delicacy of sensation by the dimensions of the retinal cones.

Not only are eyes remarkable examples of biological engineering, they are devices that have been optimized by evolution to give results that approach what is physically possible, using materials that, to a human engineer, seem far from ideal.

Chapter 3
The human eye

Optical system

The human eye is not fundamentally different in structure from the eyes of the earliest jawed fishes that evolved from chordate ancestors during the Silurian, about 430 million years ago. We know from their living relatives that the basic pattern—of a mobile globe with a lens throwing an image onto a retina of rods and cones—has survived from that period. The only important change occurred when fish came onto land about 50 million years later, during the Devonian. Then the cornea, which now separated air from fluid, became a refracting surface, and largely took over the role of image formation from the lens. Because of the power of the cornea, the lens in land vertebrates became flattened and less powerful than the spherical lenses of our aquatic ancestors. In humans about two-thirds of the ray bending is performed by the cornea and one-third by the lens. Unlike the eyes of fish, where the lens is spherical and image quality is the same in all directions, both the cornea and lens in human eyes have a single axis, which means that the image is best resolved where this axis meets the retina, and decreases in quality away from this region. The fovea, the region of highest density of receptors, is located close to this axis (Figure 12).

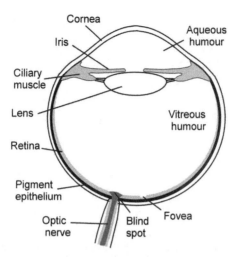

12. Diagram of a section through the human eye, showing the optical system of the cornea, lens, and iris, the retina with the central high resolution fovea, and the blind spot where the optic nerve leaves the eye

Focus

On land the lens acquired a new role: that of adjusting the power of the optical system to focus on objects at different distances (accommodation). In fish this had to be done by moving the spherical lens bodily towards or away from the retina, but in mammals and birds it is done by changing the shape of the lens. To focus on close objects the lens needs to become more powerful. The lens is elastic, at least in younger people, and when the tension on it is released it tends to bulge. This increases the curvature of the lens surfaces and so strengthens the optical power of the lens itself, enabling the eye to focus on closer objects. Somewhat paradoxically, in the unstimulated state the lens is held under tension, and this tension is released by contraction of the surrounding ciliary muscle. This muscle acts as a sphincter, which when stimulated decreases in diameter and so allows the lens to

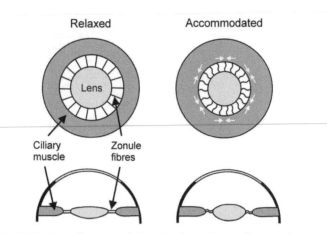

13. Mechanism of accommodation. The figure shows a front-on view (above) and side view (below) of the lens and its supporting structures. In the unstimulated state (left) the ciliary muscle is relaxed, and its inner margin expanded. This puts the zonule fibres that support the lens under tension, and the lens is stretched. During accommodation the circular muscle fibres of the ciliary muscle contract (arrows), reducing the diameter of the inner aperture. The zonule fibres are no longer under tension, the lens reduces in diameter and its front and back faces bulge. This decreases their radii of curvature, and hence increases the power of the lens

bulge (Figure 13). Activation of the ciliary muscle occurs when the retina detects blur in the image.

Common defects

The elasticity of the lens decreases steadily with age. At age 1 the minimum focus distance (the near point) is about 5 cm from the eye, increasing to 10 cm by age 20, 20 cm by age 40, and arm's length by 55. At this distance ordinary print is so far away that unaided reading is impossible. This condition is known as presbyopia, is unavoidable, and is simply cured with reading glasses. Opticians specify the strength of glasses in dioptres (or diopters), a measure of optical power, given by the reciprocal of the focal length of the lens in metres. Thus two-dioptre reading glasses have a focal length of 50 cm, which means that a book held 50 cm from the eye can be read by someone with no residual accommodative power in their eyes.

Ideally, eyes in which accommodation is relaxed are focused on the far distance, and are said to be emmetropic. This is often not the case, and for many the relaxed eye is focused on a plane not far in front of the eye (myopia, or 'short sight'). This condition can result either because the optical system of the eye is too powerful, or because the distance from lens to retina is too long; in either case the image for distant points will fall in front of the retina. Concave glasses or contact lenses (i.e. lenses with negative power) will increase the overall focal length, bringing the image back into focus (Figure 14). Again, the dioptre measure is useful. If it requires a minus-4 dioptre lens to bring the relaxed eye to a distant focus then the eye is said to have 4 dioptres of myopia; it also implies that the furthest distance that the subject can focus without this correction (the far point) will be a quarter of a metre (25 cm) from the eye. The opposite condition (hyperopia or 'long sight') occurs when the eye's own optical system is not powerful enough, or the length of the eye is too short. The image for distant objects then falls behind the retina. Accommodation can bring the image back onto the retina, but not in ageing eyes, and then

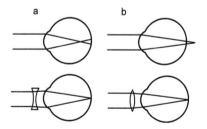

14. (a) Myopia; and (b) hyperopia. Lower figures show glasses used to bring the focus back onto the retina

positive lenses are needed to provide a focused image, even for distant objects. The other common defect is astigmatism, in which the focal length of the lens or cornea is different in different planes. This causes a point image to become a line, and blurs the image as a whole. The easiest way to cure this is to use astigmatic glasses in which the powers in different planes are complementary to those of the eye itself.

Maintenance: blinking and tears

All terrestrial vertebrates blink in order to spread fluid across the outer surface of the eye. This lubricates and cleans the cornea, and prevents it drying out. In mammals the eyelids are used, but in birds and reptiles another membrane, the nictitating membrane, moves sideways across the eye. Blinking happens spontaneously, roughly ten times a minute in adults, although less frequently— once or twice a minute—in infants. Concentration on a task, such as reading, also slows the rate to three–four blinks per minute. Blinking also occurs reflexly when irritants such as dust or smoke particles settle on the cornea.

Tears are produced from the lachrymal glands under the upper eyelid. The lachrymal fluid is for the most part a clear fluid resembling blood plasma, but it also contains oils and anti-bacterial agents such as lysozyme. The main function of tears is to lubricate the surface of the eye, which requires about 1 millilitre per day; this drains into two lachrymal canals at the inner corner of the eyelids, and ultimately into the nasal cavity. Tears are produced in greater quantity by reflex action after irritation by particles, or by gaseous substances such as the vapour from onions or tear gas. Humans also weep or cry in response to strong emotions such as grief, rage, or great happiness. This crying is initiated in parts of the limbic system of the brain, responsible for emotional behaviour, and this type of crying seems to be unique to humans. Other mammals have a variety of distress calls, particularly when very young, but they do not cry tearfully as humans do.

Iris and the pupil response

In front of the lens is the iris, a contractile ring of tissue that is opaque, and characteristically pigmented in shades of blue, green, or brown (Figure 12). The iris can close the pupil down to 2 mm in bright light, and open it to 8 mm in the dark. The change in image brightness is a factor of 16, which is insignificant when compared to the total range of 10^{10} over which the eye operates. The more likely role of the iris, rather than simply acting as a controller of image brightness, is to ensure maximum resolution under existing lighting conditions. Since the resolution improves with increasing image brightness, but becomes degraded if the aperture size goes above 3 mm, a compromise is involved which is not straightforward.

The diameter of the pupil is governed by the amount of light reaching the retina, which sends information via the brain to the dilator and sphincter muscles that open and contract the iris. This pathway is part of the autonomic nervous system, which operates unconsciously and can be affected by emotional factors such as arousal—sexual or otherwise. Famously, Roman and medieval women used to rub henbane or belladonna (containing atropine, a substance that interferes with the workings of the neurotransmitter acetylcholine) into their eyes to make their pupils bigger, which, they believed with some justification, made them look more alluring.

The human iris is unusual in the high contrast it provides with the white sclera that surrounds it; in no other primates is this so pronounced. The position of the white–dark boundary makes it possible to judge the direction of someone else's gaze with high precision, and no doubt this is related to the extreme importance we attach to eye contact in all kinds of social interaction.

The retina

The vertebrate retina arises as an outgrowth of the developing brain turned in on itself to form a cup. This curious origin accounts for the strange inverted structure of the retina, with the receptors in the deepest layer, furthest from the light. In front of the receptors, in the light path, are layers of neurons that organize the information from the receptors, and ultimately send fibres into the optic nerve (Figure 15). This arrangement is not as deleterious as it might sound, as the neural layers are transparent and cause little image distortion. This inverted arrangement of the retina is not shared by eyes in other phyla: the eyes of octopus and squid, for example, have receptors facing the light, and the processing neurons in a separate structure outside the eye.

The receptors, rods which deal with low light levels and cones that provide colour information (Chapter 6), respond to light by closing their cell membrane to positively charged sodium ions and becoming internally more negative. This voltage signal is transmitted down the receptors to the junction with the next cell layer, the bipolar cells (Figure 15). These cells link the receptors to the final stage in the retinal processing chain, the ganglion cells. The ganglion cells are situated nearest to the retinal surface, and they have long nerve fibres (axons) which collect together into the optic nerve which runs to the brain. Unlike most of the cells in the retina, which transmit simple graded voltage signals, the ganglion cells produce all-or-none action potentials, which do not decay over long neural distances. The outputs of bipolar cells are already quite complex, and are of two kinds, responding either to increases (ON) or decreases in light (OFF). Their outputs are spatially organized in such a way that the ganglion cells, for which they provide the input, respond to circular patterns (receptive fields) which either have an ON-responding centre with an OFF-responding region surrounding it, or the other way round (Figure 16).

There are two other main cell types in the retina, which spread at right angles to the 'vertical' receptor–bipolar–ganglion cell route. These are the horizontal cells which form a lateral network between the receptors and the bipolar cells, and the amacrine cells

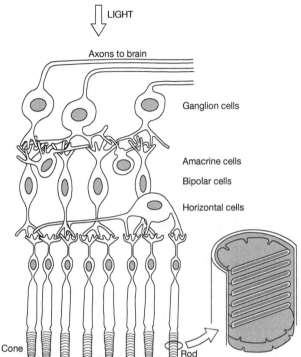

15. **Diagram of the cells in a typical vertebrate retina. The receptors are furthest from the light, and connect, via the bipolar cells, to the ganglion cells which form the output of the retina to the brain. Two layers of cells, the horizontal cells and the amacrine cells, make connexions across the retina, and modify the responses of the 'straight-through' receptor—bipolar—ganglion cell pathway. The insert shows the way that the membranes of rods, which contain the rhodopsin, are stacked into discs to increase their light-capturing area**

which form a similar network between the bipolars and ganglion cells. The horizontal cells make inhibitory connections between receptors which enhance the differences between the receptor responses, making them more sensitive to local contrast and less

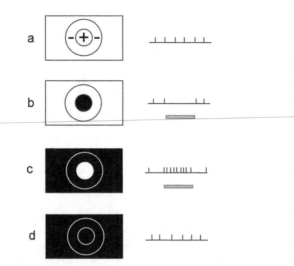

The Eye

16. **Responses of an ON-centre ganglion cell to (a) uniform light; (b) a dark spot; (c) a light spot; and (d) uniform dark. The figures on the left show four distributions of light and dark on the retina, with rings corresponding to the centre (+) and surround (−) regions of the receptive field. The figures on the right show the response measured in the axon of the ganglion cell to the patterns on the left. They show the action potentials ('spikes') produced over a period of a few seconds. Notice that the response to a uniform field—black or white—is very weak compared to the burst produced by a white spot on a dark background (c). An illuminated surround on its own inhibits the cell (b). An OFF-centre cell would respond the same to (a) and (d), but give opposite responses to (b) and (c)**

to simple intensity. The amacrine cells are of many types, and contribute, in ways that are barely understood, to the organization of the receptive fields of the ganglion cells. In view of its neural complexity, the retina is best thought of not just as a receptive structure, but as the part of the brain in which the first two stages of image processing occur.

The ganglion cells themselves are more or less unresponsive to uniform light or dark (Figure 16). What they do like are differences in illumination between the central and surround regions of their receptive fields. This means they are sensitive to the presence of detail that provides local contrast. The receptive fields come in various sizes across the retina but tend either to be small (p cells) or large (m cells). The p cells are ultimately responsible for fine discrimination. Many are wavelength sensitive and responsible for colour (see Chapter 6) and in the fovea they may have receptive fields limited to a few cones. The m cells have larger receptive fields, respond faster, and are particularly suited to detecting motion in the image.

The axons of the ganglion cells pass across the surface of the retina and collect into the optic nerve head, where they burrow through the retina and leave the eye in the optic nerve. There are about one million axons in the optic nerve, which represents a huge reduction in numbers compared to the five million cones and 100 million rods in the retina. This reduction also corresponds to a condensation of the raw information from the receptors into more succinct and useful forms. The arteries and veins to and from the retina run with the optic nerve and enter the eye at the same point. Because of all this traffic the optic nerve head takes up quite a large area, which results in a region devoid of retinal tissue. It is referred to as the blind spot, because it is insensitive to light. It is situated about 15° from the fovea and is easily detectable using the scheme in Figure 17. We do not normally see the blind spot as a blank in the field of view because the brain somehow fills in the void with material that blandly matches what is around it.

The blood vessels from the blind spot run across the retina, supplying it with oxygen. They are normally invisible, even though

+

●

17. The blind spot. Close the left eye and look at the cross with the right. As you move the page nearer and further away the black spot will disappear and reappear

they are in the light path, because, being permanent features, the retina simply adapts to their presence. They can be made visible, however, by shining a focused light onto the white of the eye from the side. This casts shadows of the vessels onto previously unstimulated receptors, and the pattern of vessels suddenly appears as a shimmering web. This quite often happens by chance during eye examinations.

Fovea and periphery

The retina is far from uniform. In the centre, around the point where we direct our gaze, is the fovea (Figure 12). This is a slight depression from which the ganglion cells have been displaced, so that the receptors have little overlying tissue and no blood vessels to impede reception of the image. The fovea has the highest density of receptors, which are all cones, and it is tiny, about 0.6 mm across. It is from this small region that we derive almost all our information about colour and detail (I will discuss the roles of different cones in colour vision in Chapter 6). Projected onto space the fovea occupies an angle of 2°—about the width of a thumbnail at arm's length, or a face at 5 metres. Resolution is at its highest here, and falls dramatically with distance from the centre, so that at 20° from the fovea (corresponding to about 6 mm on the retina) it has fallen to a tenth of the central value. This decrease occurs because the density of both receptors and ganglion cells falls off sharply. The rods, responsible for dim light vision, begin to appear towards the edge of the fovea, and are dominant through the whole of the rest of the retina, with cones becoming increasingly scarce towards the periphery. It is sometimes difficult to believe that peripheral vision is as bad as it is. Because central vision is sharp, and we move our eyes to look at anything of interest with the fovea, we get the impression that the whole field of view is detailed and colourful. But you only have to try to resolve letters in a word one or two away from the one you are looking at now to realize that this is not true.

Human eyes, and primate eyes in general, are unusual among mammals in having such a high concentration of resolution in a small central region. This no doubt reflects our early food finding strategies, and also the need for primates to recognize the facial features of others at a distance. Ungulates, such as horses, cows, and antelope, that graze on flat plains have a horizontally extended region of high ganglion cell density called a visual streak, which images the horizon. The same is true of many sea birds. Carnivores such as cats and dogs do have a centre-weighted retina, but to nothing like the same extent as in primates. Ground-feeding birds such as pigeons often have two foveas, one pointing laterally to detect food grains and one directed forwards along the beak for accurate pecking. Many fish have uniform retinas, but others that feed near flat sandy bottoms have visual streaks similar to those of ungulates. Thus, although the cell types that make up the retina are the same in all vertebrates, their numbers and layout vary with the way of life of particular species.

Chapter 4
The moving eye

Saccades and fixations

If you look round a room, the subjective impression you have is that your gaze is moving smoothly over the objects around you. During reading it seems that vision moves continuously along the line. However, watching someone *else* look around a scene, or read, gives a very different impression. The eyes jump and then stop briefly before jumping and stopping again, in a repeating pattern. Gaze does not wander, it leaps about. Late in the 19th century, when it was realized that this is what was going on, the French word *saccade* (twitch or jerk) was co-opted to describe the fast movements, with *fixations* for the stationary periods in between.

Why does our eye movement system not allow gaze to move continuously? The answer lies in the slow response time of the receptors. It takes about 20 milliseconds (1/50th of a second) for a retinal cone to respond fully to a change in intensity, so anything in the image that changes in a shorter time than this will not be fully recorded. In a moving image this means that features that take less than 20 milliseconds to pass across a cone will not be properly resolved; in angular space this represents a speed of about one degree per second, which is very slow indeed. The consequence of allowing gaze to move is motion blur, which everyone experiences looking out of a train window at nearby foliage. The effect can also be seen by tracking one's moving finger in front of a detailed scene—a row of books, for example—and looking beyond the finger at the background, which will have lost all detail. The answer to this problem is the saccade-and-fixate strategy, which all vertebrates share. Information is taken in during the fixations, and saccades move the eyes as fast as possible to change the fixation point. Typically, in humans, a saccade lasts about 30 milliseconds and each fixation about 300 milliseconds, so that we make about three fixations a second during waking hours. It also means that if we add up the saccades, during which we are effectively blind, this comes to about an hour and a half of visual 'down time' per day.

Stabilizing the image

When viewing the world we do not just move our eyes. In fact changes of gaze direction (the direction that the eyes are pointing *in space* rather than in the head) are performed predominantly by movements of the head. But since the eyes are attached to the head this means that, other things being equal, the eyes will be dragged round by the moving head, and vision will be compromised by the image movement. This problem must first have arisen in early fish, which made turns and undulations while swimming, all of which would cause the image of the surroundings to move across the retina in an unhelpful way. The solution that evolved was to use the semi-circular canals of the inner ear to measure the rotation of the head, and to take advantage of this signal to move the eyes in the opposite direction. In this way gaze direction could remain constant, because the eye-in-space direction (gaze) is equal to the eye-in-head direction plus the head-in-space direction. This ubiquitous reflex is known as the vestibular-ocular reflex, or VOR. An example of how this works in practice is shown in Figure 18, which is a recording of the author's eye, head, and gaze movements during three seconds of looking round a room.

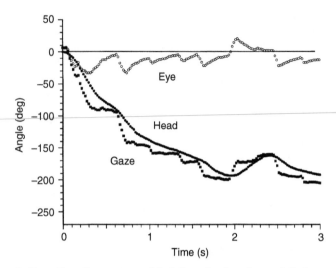

18. Recording of movements of the left eye, head, and gaze while the author looked round his kitchen. The upper record shows the movements of the eye's axis relative to the direction of the head. It makes a series of saccades, and between them it rotates back to the forward direction at a speed opposite to that of the head (VOR). The head (solid dots) moves more slowly in the same direction as the eye saccades, but ultimately causes nearly all the change in gaze direction. The gaze record, the movement of the eye's axis in space, is the sum of the eye and head rotations. It contains the eye saccades but, because of the VOR, gaze is almost motionless during the fixations between saccades

In the recording it can be seen that most of the movement, through about 180°, is made by the head. Gaze moves with the head, but with jerky saccades interspersed with stable periods when little movement occurs—these are the fixations. The eyes, which have limited movement in their sockets, make a series of rapid saccades, but return each time to the straight ahead position. During these returns they move in the opposite direction to the head, but at the same speed. This is the VOR in action, and the sum of the eye and head movements gives rise to the characteristic stepped pattern of gaze movement.

Other eye movements

There are only four kinds of eye movement. In addition to saccades and the slow movements that compensate for head movements there are two others: pursuit and vergence movements (Figure 19). You can follow a moving object, for instance a football, quite smoothly, provided it does not go too fast. If it moves faster than about 15 degrees per second in the field of view you can still track it, but the eye movements will become increasingly saccadic. Interestingly, smooth tracking can only occur if there is a real object to track; you cannot track an imaginary object, other than by making saccades. Vergence movements are made when a fixated object moves in depth. These movements differ from all the others in that the eyes move in opposite directions. For distant objects the eye axes are parallel, but as the object approaches the axes converge, crossing where they meet the target.

Blinks are not strictly eye movements, but like saccades they interrupt vision for periods of 0.1 to 0.3 seconds. They often accompany large saccades, presumably to synchronize the two sources of lost vision. Spontaneous blinks occur about 12 times a

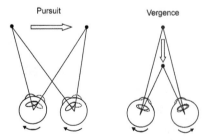

19. Pursuit and vergence eye movements. Solid and dashed outlines show the eye orientation at the beginning and end of the movement. In pursuit, and during saccades and compensatory movements, both eyes rotate in the same direction. Only during vergence, when the object moves in depth, do the eyes rotate in opposite directions

minute, during which the eyelids sweep tears downwards and medially across the cornea, cleaning its surface. Protective reflex blinking can also be triggered by objects touching the cornea, objects seen approaching, or by sudden light or sounds.

Eye movement machinery

Each eye is moved by six muscles (Figure 20), which both rotate the eye and help to keep it suspended in the orbit. The four rectus muscles move the eye axis vertically and sideways, and it is these muscles that make it possible to target different points in the surroundings. The other two oblique muscles rotate the eyeball around its axis, and allow some degree of compensation for head roll. The muscles are organized in opposing pairs which means that if one muscle contracts, its opposing muscle (antagonist) must relax. The neural structures that innervate the muscles and manage the complexities of their interplay are the three pairs of oculomotor nuclei in the base of brain (Figure 21). These nuclei do not decide where to look, but they translate commands, ultimately from the cerebral cortex, into patterns of activation that will turn these

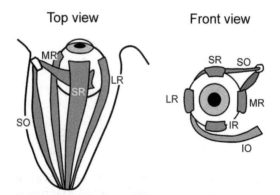

20. The arrangement of eye muscles in the orbit. LR, IR, MR, and SR: lateral, inferior, medial, and superior rectus muscles. IO and SO: inferior and superior oblique muscles

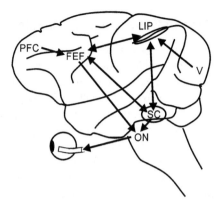

21. Outline of the brain structures involved in the control of eye movements. FEF, frontal eye fields; LIP, lateral intra-parietal area; ON, oculomotor nuclei; PFC, pre-frontal cortex; SC, superior colliculus; V, visual cortex. See also Chapter 7

instructions into directed eye movements. The immediate control centres for these nuclei are the two superior colliculi in the mid-brain. They coordinate the competing calls on the eye muscle machinery from different parts of the cortex and elsewhere in the brain. The colliculi also apportion how much of a gaze movement will be performed by the eyes and how much by the neck muscles that turn the head. The main cortical regions controlling eye movements are the left and right frontal eye fields, and a small region of the parietal lobes known as the lateral intra-parietal area. These are the sites where commands for 'voluntary' gaze shifts are initiated, and they can produce eye movements either via the superior colliculi, or instruct the oculomotor nuclei directly. These control structures are themselves influenced by regions such as the pre-frontal cortex that control the tasks that the eyes must perform. This three-level control system—cortex, colliculi, and oculomotor nuclei—allows for varying levels of control, from the reflex orientation which occurs when an intruder appears in peripheral vision, to the intricate viewing strategies of an artist working on a portrait.

Patterns of eye movement in everyday life

What determines where we look next? This innocent-looking question has been debated for most of the last century, and remains largely unresolved. Sometimes the answer is obvious: the sudden appearance of a novel object will nearly always trigger a saccade. But this is a rare event, and most of our interactions with the world are not like this. In general we use our eyes to gather information that we need, either to answer a question that is currently in our mind, or to aid in the control of actions. One of the first people to document this was the Russian physiologist Alfred Yarbus. During the 1950s Yarbus developed a technique of recording eye movements by bouncing a beam of light off a mirror attached to a contact lens that moved with the eye. The beam then wrote directly onto a photographic plate. He then asked his subjects different questions while looking at a picture. In Figure 22 this picture is (a) 'An Unexpected Visitor' by I. P. Repin. He found that each question evoked a different pattern of eye movement. In the figure the dense 'knots' are fixations and the lines between are the saccades that link them. In (b) three minutes of free viewing covers most of the scene, but with a clear emphasis on the faces of the participants. In (c) the subject scrutinizes the clothes, as expected from the instruction. Particularly interesting is (d); the question is barely answerable but relationships are the key, and the subject keeps looking from face to face in search of clues. Interestingly the attitude of the child on the right is important here, but the housekeeper, who is presumably not family, is ignored. What Yarbus has demonstrated here is that most eye movements are not reflexive, but are governed by a plan for moving the fovea around to those parts of the scene likely to yield the most relevant information.

Eye movements and actions

In many activities it takes time to process visual information before it can be converted to motor behaviour. As early as 1920,

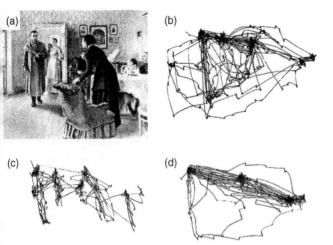

22. Eye movement recordings by Alfred Yarbus of a subject viewing a picture (a) 'An Unexpected Visitor'; (b) free examination; (c) 'remember the clothes worn by the people'; (d) 'estimate how long the unexpected visitor had been away from the family'. Each recording lasts 3 minutes

George Buswell showed that when reading aloud the location of gaze on a printed line is about one second ahead of the spoken word. A similar one-second time interval is found in copy-typing, and in sight-reading music when playing the piano. That there is a delay is not surprising. In all these activities much has to happen between the acquiring of information and the production of a response. In music reading for example, notes are read in small clusters, alternating between the upper and lower halves of the stave. These have to be decoded and assembled into instructions for the fingers, and the notes then have to be played in a smooth sequence, with both hands at the same time. As H. E. Weaver, who made the first such study in 1943, put it:

Notes on the treble and bass parts of the great staff are usually so far apart that both vertical and horizontal movements of the eyes must be used in preparing two parallel lines of material for a unified performance.

In all these activities, the visual functions, eye movement and motor regions of the brain, have to act together to produce what is effectively a continuous production line, with information acquired in a series of disjointed fixations, assembled into a continuum, converted into action, and then erased. Once the skill is acquired the process proceeds smoothly, although in the early stages of learning it seems more like a series of one-off events.

This one-second delay between fixation and subsequent action seems to be a feature of many other activities. When walking over rough terrain the eyes are about two steps ahead of the footfall they direct, which corresponds to about a second in time. In driving on a winding road the part of the lane edge that is used to direct steering is typically slightly less than a second in front of the car. Even in a complex activity such as making a cup of tea or a sandwich, where every act involves a different object, gaze is typically a half to one second ahead of each motor action, seeking the information needed for its execution. In general, then, the eye movement system directs gaze in a prospective way, in advance of action, rather than simply checking retrospectively that the action has been properly performed.

The need for advance information is particularly acute in sport. It takes one-tenth of a second or more for information in the image to reach the cortex. This is a long time in a game of tennis, and it means that controlling fast action simply on the basis of moment-by-moment visual feedback is impossible. The key to this is anticipation: to gain information about what is going to happen before it happens.

A good example of this is shown in Figure 23. The picture on the left is a photograph, taken in 1902, of the great batsman Victor Trumper, preparing to make a straight drive at the Oval cricket

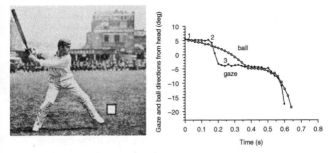

23. Left: a batsman about to strike a ball in cricket. The ball must occupy the small square at the moment of the strike. Right: record of the gaze movements of a batsman before, during, and after the bounce of the ball. The batsman makes a downward saccade at (2), about 0.14 seconds after the ball has emerged from the bowling machine. This saccade results in a fixation close to where the ball will bounce (the inflection in the ball record after 0.37 seconds)

ground. The accuracy of timing required in fast ball sports is extraordinary. In cricket the ball usually bounces partway down the pitch, and much of the information needed to get the timing of the stroke right can be obtained from the place of the bounce; batsmen need to make sure they acquire this information. On the right of Figure 23 is a recording of the vertical gaze direction of a batsman, relative to his head, as he observes the flight of the ball from the time it leaves the bowler's hand (1) to the moment of impact. The most striking feature is that he does not 'keep his eye on the ball' (which is what coaches tell you to do), but makes a saccade (2) to the point where he expects the ball to land, and waits there (3) until it does. Thereafter he tracks the ball until it reaches the bat. With this strategy he has time to observe the time and location of the bounce point, and adjust his stroke accordingly. If he had simply followed the ball to the bounce he would have been too late, because starting smooth pursuit takes about 0.2 seconds, and even then the eye continues to lag behind the ball. This is much longer than the 0.14 seconds it took to start the saccade in the record. Anticipatory eye movements of this kind have been recorded in table tennis and in squash, and are probably universal in ball sports.

Do we know where we are looking?

We often classify actions as 'voluntary' or 'reflex', but this doesn't really capture the essence of what is happening. It is true that we can decide to direct our gaze to some particular feature—just as we can consciously control our breathing—but we rarely intrude in this way into a process that normally proceeds, unsupervised, according to its own rules. The batsmen in the last section did not know they made a saccade that took their eyes to the bounce point, and motorists are unlikely to be able to tell you which road features they looked at to judge how much to turn the steering wheel. The eye movement system has its own rules, and its own knowledge base, and 'we' are rarely in charge. Fixation strategies, it seems, are learned and elaborated, piecemeal and without instruction, from birth through into adulthood. Very few instruction manuals, whether concerned with DIY, piano playing, driving, or sport, actually tell you where to look. This is quite surprising, given that the crucial role of the eye movement system is to know where to look to get the information that will allow the action to proceed. As pointed out earlier, many parts of the brain are involved in making eye movements and deciding where to look, and there is no single control centre that directs gaze moment by moment. At present we have no coherent theory of gaze control.

Eye movements of other animals

The problem of slow receptor response time applies to all eyes, and so it is not surprising that a saccade and fixation strategy of one sort or another occurs in almost all animals with good eyesight. This is true of all vertebrates, many insects and crustaceans, and cephalopod molluscs such as cuttlefish. However, there are differences in the way it is executed. Insects have eyes that are fixed parts of the head, so a fly in flight makes neck movements where we would make eye movements. Flies have a reflex, analogous to our vestibular ocular reflex but based on

50

balance organs known as 'halteres', which produces counter-rotation of the head when the body rotates, thus stabilizing gaze direction. Hoverflies are so agile in flight that they do not even make neck movements when manoeuvring, but make saccade-like turns using the wings alone to rotate body and head together. Unlike most other vertebrates, birds make quick head jerks in place of eye saccades when looking around, which makes them appear particularly alert. Presumably they can do this because their heads are so light that they can move them as fast as we can move our eyes. Ground feeding birds, such as chickens and pigeons, have another characteristic behaviour related to vision, often referred to as head-bobbing. When walking they thrust the head forward, then keep it still in space as the body walks under it, before thrusting it forward again. This seems to be a way of maintaining a clear view to the side, while still allowing the bird to move forwards.

Chapter 5
The third dimension

Each retina has a two-dimensional image of the world, akin to a photograph or a movie projected onto a screen. As in a photograph, there will be information about relative distances that can be inferred from the sizes and locations of objects in the picture, but there is no representation of the third dimension in the flat surface itself. Yet even with only one eye the sense that the world has depth is utterly persuasive. With two eyes there *is* explicit depth information, available from the small differences between the images in the two eyes. Primates like ourselves make great use of this information for manipulating objects in the space immediately around us. But even for a rabbit, with eyes pointing sideways and almost no overlap between the fields of view of the two eyes, there are implicit cues to distance that are adequate, at least for locomotion and social interaction. Thus there are two rather distinct kinds of information available for the construction of the distance dimension of the perceived world: binocular cues, often referred to as stereopsis, and monocular cues, from the pictorial content of the image and the motion of parts of the image when the head and body move.

Binocular stereopsis

If you hold up a finger from each hand in front of you at slightly different distances, and alternately close one eye and then the other, the relative positions of the fingers will change dramatically because each eye has a different viewpoint. These shifts are by far the strongest cues to distances up to a few metres away. If you try the same thing with distant trees it doesn't work, because the distances involved are large compared to the distance between the eyes. The geometry involved is shown in Figure 24.

Retinal disparity is the difference in retinal distance, relative to the centres of the two foveas, of the images of objects in different distance planes. In Figure 24, where the two black objects are at

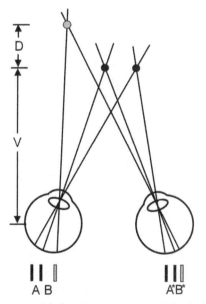

24. **Disparity. Left and right eyes view three vertical rods, with the two foveas centred on the middle rod. The right-hand black rod is at the same distance as the middle rod, but the grey rod is further away. Shown beneath the eyes are the image positions on the two retinas. Distances A and A* are the same, but B and B* are different. B–B* is the retinal disparity of the grey rod**

the same distance from the eye, the two retinal distances A and A* are the same, and the disparity is zero. For the grey rod, however, the retinal distances B and B* are different, and the disparity is B–B*. If the grey rod were closer than the black rods the retinal distances in the two eyes would be reversed, and the disparity would have the opposite sign. It is clear that in 95 per cent of the human population the brain can compute both the magnitude and sign of these disparities, and convert them to relative distance. The word relative is important here. The retinal disparity depends on both the depth difference (D on the Figure) and the absolute viewing distance V: in fact it is proportional to D/V^2. This means that neither D nor V is knowable independently of the other.

There is, however, a direct measure of viewing distance (V) that the eyes can supply. When focusing on a near object (such as the central black rod in Figure 24) the two eyes have to converge, and the angle between their axes gives a simple and reasonably accurate measure of V. This information is available from the differences in the neural commands to the eye muscles, when the eyes are appropriately converged. This, in turn, can be used to calibrate the relative distance information supplied by disparities in the image, and so provide absolute estimates of depth. Without this clever bit of geometric computation, threading needles and mending watches would be a great deal harder.

The discoverer of the importance of disparity was Charles Wheatstone in 1838. He realized that it was the differences in the images in the two eyes that made stereoscopic depth vision possible, and he went on to invent a stereoscope in which two pictures from different viewpoints could be presented separately to the two eyes using mirrors. A later version of the stereoscope using prisms became very popular in Victorian times, especially when allied to photography. Later versions involving red-green or Polaroid glasses all use a similar principle.

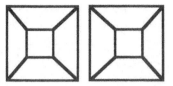

25. Simple stereogram showing how images from different viewpoints can give rise to depth. Squint slightly so that the two images combine to give a single central square and two outer ones. The central square should contain a flat-topped pyramid pointing out of the page. The 'magic eye' technique will also work, but the depth will be reversed

It is also possible to view stereo pairs unaided. In a stereo pair like that in Figure 25 the trick is to get each eye to see just one of the two images. There are two ways to do this. One is to squint, so that the left eye views the right-hand image and the right eye views the left-hand image. You should then finish up with three squares, the central one of which is the fused image from the two eyes. This image will be seen in depth as a topless pyramid emerging from the page. It may help to put your finger halfway between your eyes and the page and focus on that temporarily. It takes a little while to learn the art of squinting by the right amount, but once this is achieved the squares will snap into place and disparity will do the rest. You will then have this skill for life. Alternatively you can use the 'magic eye' technique, in which the right eye views the right image and the left eye the left image. The best way to do this is to let your eyes relax so that they are looking into the distance, and then bring the pictures into focus without letting the eyes converge onto the plane of the page. You will again have three images, but in this case the central image will have an inset square that is further away, so that the picture looks like a square wall at the end of a corridor. Either method requires a little practice, but the impression of depth is worth the effort. The reason that this is difficult is that focus and convergence are normally coupled, and here you have

26. Random dot stereogram of the type devised by Bela Julesz. Fuse the images in the same way as in Figure 25. A small square will appear about a centimetre in front of the larger square (or behind if using the 'magic eye' method). It may take a few seconds to emerge. This shows that disparity can operate on textures as well as objects

to de-couple them to get illusion to work; in a proper stereoscope this de-coupling is done for you with prisms and lenses.

It turns out that it is not necessary to have recognizable objects in the scene for the disparity mechanism to operate. In the 1960s Bela Julesz produced the random-dot stereogram, in which fields of apparently random small square dots were presented to the two eyes. In the classic example most of the two fields are the same, but an undetectable central square of dots has been shifted by a few dots, in different directions in the two fields. The result is a central square that stands out in depth (Figure 26).

Stereograms of the Julesz type have been very useful in exploring the limits of the disparity mechanism, as well as in diagnosing problems when it fails to function. Sensitivity to disparity is extraordinary, and the resolution of the system is at least as good as that of the foveal receptor mosaic, where, in angular terms, the cone separation is about 0.5 arc minutes. This implies an ability to detect a depth difference of 1 mm at a distance of 7 metres. Not everyone shares this ability. A small proportion of the population, between 1 and 5 per cent, lacks the mechanism for combining the images from the two eyes. Often this arises when a child's eyes are misaligned by a squint, which leads to double vision (diplopia). In such cases the brain tends to suppress one image, leading to a condition known as amblyopia, or, more commonly, 'lazy eye'. If this occurs the connections in the cortex that are needed to set up the disparity mechanism do not develop, and the child is left totally or partially stereo-blind. Cures include dealing with the squint surgically, or using prismatic glasses to bring the two images together. Both work if done early enough.

The exact brain mechanisms involved in disparity detection are still a subject of debate. The prerequisite is that there should be cells in the cortical visual areas which receive inputs from both eyes, but which have slightly different receptive field locations in each eye. These cells will then pick up the same image features, but at different disparities. Such cells have been found in area V_1, which receives nearly all the visual input to the cortex, and also in neighbouring areas V_2, V_{3a}, and V_4 (see Chapter 7). Whether one, all, or none of these regions contribute to the subjective appearance of the three-dimensional world around us goes to the heart of the question of how the brain produces the image we see.

Pictorial cues

For distances beyond about 10 metres the binocular mechanisms that actually measure depth become less effective and we have to rely increasingly on the other kinds of cue provided by the content of the scene. There are many of them. During the Renaissance painters discovered how to use these cues to give their pictures depth, so that although essentially flat they convey a three-dimensional world with impressive realism. Rather later, Canaletto (1697–1786) became a master of these techniques in his handling of huge scenes, particularly cityscapes, and I have used one of his paintings of 'The Grand Canal in Venice' (Figure 27) to illustrate some of the many features of a scene that provide depth information.

27. *The Grand Canal,* Venice, by Canaletto, about 1740

Relative size. The figures in the foreground are larger in their angular extent than those in the background. Since we have a feeling for the heights of people, and other objects, a direct measure of distance is given by their estimated height divided by their angular size.

Perspective. A discovery of the early Italian Renaissance was that lines parallel to the viewer's line of sight converge to a 'vanishing point' on the horizon. The degree of this convergence is a strong cue to distance. The buildings on the right provide a variety of such lines. Notice that the slight distortion of the perspective lines on the more distant buildings tells us that the canal is bending to the right. Canaletto has chosen a vanishing point that is to the right of centre, and about 3 metres above the level of the nearest boats.

Texture. The water has a texture of ripples that become smaller and sparser with distance. If the water were tiled the pattern of elements would converge with distance, following the same rules of perspective as the buildings. Texture and shadow can also be useful guides to the shapes of objects, for example in the costumes of the figures.

Elevation. Because the viewer's eye level is above the ground plane, the more distant boats and figures are higher in the picture than those that are nearer. In a real scene the same would also be true: the angle down from the horizon to the feet of figures is an inverse measure of distance.

Occlusion. If one object obscures another then it must be in front of it. This obvious fact allows us to say that the masts of the boats in this picture are nearer than the buildings. This doesn't give a measure of distance, but it does provide an unambiguous order of distance.

This is not a complete list, but it does cover the more commonly available cues. One could add that in the far distance colours become bluer and contours less distinct, providing what is known as aerial perspective. But Canaletto preferred bright days and medium distance subjects. Many Impressionists did however make extensive use of such cues, as did J. M. W. Turner, and German Romantics such as Caspar David Friedrich.

Distance from motion

When you move through the world, close objects move faster across the retina than distant objects. This observation was developed by James Gibson in 1950 in an influential book *The Perception of the Visual World* in which he described the pattern of motion of objects in the field of view. During the Second World War Gibson had been concerned with the question of how pilots land aircraft, and came to the conclusion that their main source of information was the apparent motion of objects and texture elements on the ground. This has become known as the 'velocity flow field'. This pattern of retinal motion, which results from one's own locomotion, has many useful properties, but in particular it

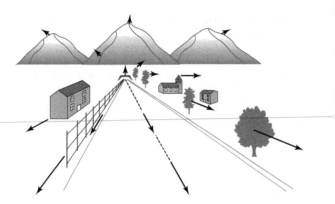

28. The velocity flow field seen by a driver on a straight road

provides another powerful cue to distance. An example of a flow field is shown in Figure 28.

The flow field has three key features. First, there is a distant point on the road ahead from which the velocity arrows radiate, and where the velocity is zero. This point indicates the current heading direction, and is sometimes referred to as the pole of the flow field. Second, velocities (shown by the arrow lengths) decrease, as the distance from the viewer increases. Third, velocities increase as the radial distance from the pole increases. This distance can be thought of as an angle in visual space, relative to the viewer's forward direction. Specifically the velocity increases with the sine of this angle. Thus objects in the direction of travel will not move at all (sin 0° = 0) and those to the side, at right angles to the line of travel, will have maximum velocity (sin 90° = 1). It might seem difficult to disentangle the effect on seen velocity of, on the one hand, the distance of objects, and on the other their angle relative to the viewer's heading. However, a walk round a furnished room, with one eye closed, usually provides a convincing demonstration that the seen motion of objects is a good guide to their distance. We seem to factor in the direction of objects relative to our line of travel, and compensate for it to obtain their true distances.

There are other, more subtle, features of the flow field that are important to us. For example, if you are approaching a surface such as a wall, the rate of expansion of the resulting flow pattern tells you, with no other assumptions, your time-to-contact with the surface. This curious fact appears to have been first noticed by the astronomer Fred Hoyle in his 1957 novel *The Black Cloud*, where he works out the time-to-contact with the eponymous cloud from its expansion rate. In ordinary life our main use of this information is probably while driving, where we use expansion based time-to-contact information as an input to a feedback loop that enables us to adjust our speed in relation to the car in front. Related to this is 'looming'. When a small part of the flow field starts to expand unexpectedly this means that an object is approaching, and the symmetry and motion of the expansion pattern indicates whether it will hit or miss. This cue is used across the animal kingdom as a warning signal, and flies will take off, locusts will jump, and humans will duck when a looming stimulus appears.

Finally, we can contrive motion in the flow field in order to estimate distance. In a situation where we need to know the relative distances of objects, beyond the limits of our stereoscopic mechanism, we often move our heads from side to side. This induces a distance dependent motion of the image in both eyes, referred to as motion parallax, which we can use either directly to estimate distance or to induce occlusions that specify which objects are in front of others. Locusts and praying mantids do exactly this when deciding to jump from stem to stem, and gerbils do something similar, but with vertical rather than horizontal head bobs.

Where is it?

We have seen that there are three very different kinds of cue to distance: stereopsis, image content, and motion. How do they come together to give the convincing, whole, image of the surroundings that we experience? Neurophysiology and brain scanning have identified a number of brain regions, mainly in the cortex, that are likely to be involved (see Chapter 7). But the regions responsible for stereo depth are not the same as those that deal with flow fields, and there are few clues as to where pictorial cues are processed. If there were a single cortical region which contained a full three-dimensional image of the visual surroundings we should know about it by now. Either it has been missed, or somehow that image is a collective product of most of the rear half of the cortex. Vertebrates other than mammals do not have a cortex like ours, and yet have no difficulty dealing with distance. In birds, undoubted masters of the third dimension, the major visual centres are greatly enlarged optic tecta. These are mid-brain structures, which in primates correspond to the superior colliculi, where they have the reduced function of coordinating eye movements. Perhaps we should not think exclusively in terms of the cortex, when trying to understand spatial vision.

The other difficult problem, which has bothered philosophers, psychologists, and physiologists for decades if not centuries, is how the brain manages to externalize its image: to put the world back 'out there'. It is almost as though it assembles the information and then projects it back out, but that makes no physical sense. With our present state of knowledge, it is not clear that we even know how to state the problem.

Chapter 6
Colour

Colour is a product of the brain

Jonathan I was a New York artist. One day he had a traffic accident that left him with a lesion in both sides of the cortex; this was not in the primary visual area V_1 (see Chapter 7), but in parts of the region surrounding it. His world went monochrome. His accounts of his condition are striking and sad. They were recounted to Oliver Sacks and Robert Wasserman and published in *The Case of the Colorblind Painter* (1987).

> He saw people's flesh, his wife's flesh, his own flesh, as an abhorrent gray: 'flesh-colored' now appeared 'rat-colored' to him. That was so even when he closed his eyes, for his preternaturally vivid ('eidetic') visual imagery was preserved but now without colour, and forced on him images, forced him to 'see' but internally with the wrongness of his achromatopsia. He found foods disgusting in their grayish, dead appearance, and had to close his eyes to eat. But this did not help very much, for the mental image of a tomato was as black as its appearance.

Mr I learned to live with his changed vision, in part by becoming nocturnal, but colour never returned. His condition—cerebral achromatopsia—is rare, and unrelated to ordinary colour blindness: in fact his eyes were entirely normal. What the passage shows is that colour is a construct of the brain, based on information supplied by the eyes, and when the region that forms that construct is gone, so is the ability to see, remember, or imagine colour.

Colour and wavelength

Objects in the world reflect different wavelengths of light to different extents, and our eyes are able to detect these differences and relay them to the brain. Humans are sensitive to a small part of the electromagnetic spectrum in the wavelength range 400–750 nanometres (nm; 1 nanometre is a billionth, or 10^{-9}, of a metre). Light that appears blue has short wavelengths (400–500 nm) and red light long wavelengths (600–750 nm) with green light in between. White light has a roughly equal amount of energy at all wavelengths. There is no sense in which the photons with these wavelengths are actually coloured: they are simply particles with energies that correspond to these wavelengths.

Figure 29 shows the reflected spectra of various coloured flowers and a leaf. Note that the colour names are the colours as they appear to us: we have no way of knowing what the brain of another animal, for example a bee, might make of them. The anemone is white because it reflects most of the spectrum. The yellow *Hypericum* (St. John's Wort) has a similar spectrum, but does not reflect in the blue region. The red *Pelargonium* is similar again, but the cut-off is now much further into the long wavelength range, so that only wavelengths above 580 nm are reflected. Both the *Hypericum* and the *Pelargonium* have secondary reflections below 400 nm, in the ultra-violet which we cannot see (but bees can). This is typical of the anthocyanin pigments of many flowers. The blue *Lobelia* also has this two-peaked spectrum, but in this case it is all shifted so far to the right that the main peak is deep into the red, and almost invisible, and it is the blue secondary peak that we actually see. The green leaf reflects in the green, but weakly—the job of chlorophyll is to absorb light rather than reflect it.

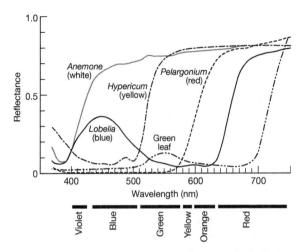

Colour

29. Proportion of light reflected by various flowers and a leaf. The ordinate is wavelength, with the corresponding spectral colours shown below. Ultra-violet wavelengths, which many insects and birds can see, lie below 400 nm on the wavelength scale

Not all colours are produced by pigments. There are also 'structural colours' produced by the interference of light waves reflected from multiple thin plates, rods, or spheres in the outer layers of the structure itself. The brilliant blue of the wings of *Morpho* butterflies, the feathers in peacocks' tails, and the green reflector behind the retina in the eyes of a cat are all examples of this kind of colour.

Trichromacy and the cones of the retina

We can see in colour because we have three different types of cone cells in the retina which respond to different parts of the spectrum (Figure 30). The idea of primary colours developed, mainly through paint mixing, in the 18th century. However, it was Thomas Young who first provided evidence that our eyes employ separate mechanisms to divide up the spectrum. In 1802 he

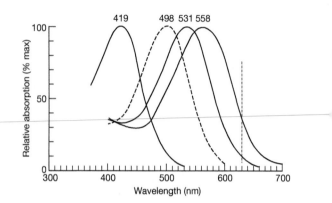

30. Spectral absorption of the three cone types and rods (dotted). Numbers give the wavelengths of maximum absorption. The dashed line (right) indicates how the cones would be stimulated by red light

concluded that we needed only three different detectors to see all the colours that we can distinguish. Young's reasons for believing this were based on matching coloured lights. For example, many spectral colours can be matched by mixtures of other spectral colours. Thus a subjective yellow can be produced by a pure 580 nm wavelength light, or by a suitable mixture of green (530 nm) and red (650 nm) lights; it was thus not necessary to have a specific 'yellow' detector. Using similar logic, Young arrived at three as the minimum number of 'primary' colours, from combinations of which all other colours could be made. In 1850 Hermann von Helmholtz made Young's argument more specific by proposing that the three mechanisms were blue-sensitive, green-sensitive, and red-sensitive. It is this system that remains the basis of RGB video and colour television systems today.

I will refer to these three mechanisms as short, middle, and long-wavelength sensitive (S, M, and L, respectively) cone mechanisms. In humans the rods, which operate in dim light and are maximally sensitive to blue-green light, are not involved in the colour vision system.

The retinal cones were first described, and distinguished from rods, by Max Schultze in 1866, but he had no way of distinguishing the three cone types. This was not possible until more than a century later. In 1983 Herbert Dartnall, James Bowmaker, and John Mollon were able to measure the absorption spectra of the visual pigments in the cones of the human retina, using eyes removed from patients because of malignant growths. They employed a technique known as micro-spectrophotometry, which involves shining a dim beam through a single cone, and determining the proportion of light absorbed at each wavelength. This is not easy, partly because it has to be done in the dark, partly because cones are only 2.5 micrometres wide (about 5 wavelengths of light), and partly because the cones bleach as the measurements are made. However, their results were clear. The visual pigment of the S cone had its maximum absorptions at 419 nm (blue-violet), the M cone at 531 nm (apple-green), and the L cone at 558 nm (yellow-green). Not red: yellow-green! The psychophysicists who had spent many years refining the Young-Helmholtz model may not have been surprised by this result, but everyone else was. It seemed outrageous that red, the primary colour that represents the long wavelength end of the spectrum, the colour of blood and danger, did not have its own visual pigment. The spectral absorption curves in Figure 30 show that when we refer to something as red, this means that the L (558 nm) cones are responding fairly well but the M (531 nm) cones hardly at all. That's red: so be it.

The main reason why our mid-wavelength and long-wavelength cones have their absorption peaks so close together in the spectrum is to do with their evolutionary history. Based on the eyes of their modern relatives, the lampreys, the first fishes had five different visual pigments in four types of cone and one type of rod. This means that their colour vision was tetrachromatic (four colour). Bony fish, reptiles, and birds have retained the full complement, but most mammals have lost two of the four pigment types, and so are left with one short-wavelength and one long-wavelength cone type: effectively they are blue-yellow dichromats (two colour). This reduction probably occurred during the age of dinosaurs, when early mammals went underground or became nocturnal to avoid trouble. However, after the extinction of the dinosaurs, 65 million years ago, the mammals flourished and radiated. About 35 million years ago, in the old world monkeys, the primate group of which we are members, the gene for the long-wavelength pigment duplicated. The duplicate genes diverged over time to give rise to two photopigments (L and M) with slightly different spectral absorptions. In this way our branch of the primates became trichromats (three colour). It seems that the new red-green channel was particularly useful to monkeys whose diet was of fruit of varying ripeness, and leaves of different ages.

Colour blindness

Both genes derived from the original L cone gene are on the X-chromosome. This is a sex chromosome, and females have two copies but males only one because it is paired with the Y-chromosome; the Y-chromosome does not carry the L and M genes. Thus in females, but not in males, a defect of one or the other gene can be supplemented by the gene from the other X-chromosome. Red-green colour blindness, or deficiency, results from a defect in one of these two genes, and it is much more common in males who have no spare copy. In Caucasian males the figure is 8 per cent of the population, compared to 0.5 per cent in females. If the L pigment is defective the condition is described as *protanopia*; if it is the M pigment it is called *deuteranopia*; and if it is the S pigment it is *tritanopia*. Tritanopia is very rare, because the S gene is on chromosome 7, and everyone has two copies. Sometimes a gene can be slightly altered, but still functional, in which case there are changes in colour vision that show up with appropriate testing, but usually cause little impairment. Some females, while having normal colour vision themselves, can be carriers of red-green colour blindness because they have one defective copy of a gene which they can then pass to their male offspring. An extreme form of colour blindness is found in *rod monochromats*: people who fail to develop cones at all. They have no fovea, no colour vision, and poor resolution. They are, however, very popular with perceptual psychologists.

Colour processing by the retina and brain

The cones supply information about wavelength in line with their spectral sensitivities, which are much the same as the absorbance spectra in Figure 30. Light of a particular hue will stimulate the various cone types to different extents, and it is the ratios between the different cone outputs that ultimately determine what colours are seen. We can see from Figure 30 that a blue-green spectral hue of 500 nm will stimulate the S cones by about 10 per cent, the M cones by 70 per cent, and the L cones by 50 per cent. This ratio defines the hue. If light levels decrease, so that these figures are halved, the ratios will stay the same, and so the hue will not change. The need for comparisons between receptor types becomes obvious if we consider the output of the L (558 nm) cone on its own. A 50 per cent response could be produced by a 500 nm light, a 615 nm light of the same intensity, or a 558 nm light of half that intensity, or by many other combinations of wavelength and intensity. Colour vision with one type of receptor is impossible because wavelength and intensity are confounded.

Colour information does not leave the eye in the form of simple responses from the S, M, and L cones, but as differences between them. The ganglion cells that carry colour information to the brain have a 'colour opponent' structure, responding, for example, positively to red and negatively to green, or vice versa. There are just two colour channels, corresponding approximately to red (L) versus green (M) and blue (S) versus yellow (L + M) cone responses, with a third 'luminance' channel formed from the summed responses of the red (L) and green (M) cones (Figure 31). Interestingly, an opponent system like this was first proposed by

Ewald Hering in 1892, who noticed that some colour pairs are never seen as combinations: for example, red and green do not give reddish-green but produce yellow, and blue and yellow do not give yellowish-blue, but green. Based on this and other evidence, Hering proposed that there were two opponent colour axes, red-green and blue-yellow. It seemed for some time that Hering's theory and the Young-Helmholtz theory were rivals, but in fact both were right: they both operate, but at different stages in the colour pathway. The Young-Helmholtz theory describes the behaviour of cones and Hering's opponent theory the subsequent behaviour of retinal ganglion cells. They are related to each other as shown in Figure 31. It is intriguing that Hering's blue-yellow pathway probably represents the old mammalian dichromatic system, while the red-green pathway is a 'modern' primate invention.

Opponent responses, although useful in demarcating boundaries in the image on the basis of wavelength differences, do not themselves provide pure hues. As we have seen from the testimony of Jonathan I and other patients with cerebral achromatopsia, subjective colour must be derived from these opponent responses further up the neural pathway than area V1 where the input from the eye arrives in the cortex (see Chapter 7). Recent evidence from scanning studies in humans and single electrode studies in monkeys suggests that narrowly defined hues first arise in the posterior inferior temporal cortex, two neural stages beyond V1, in

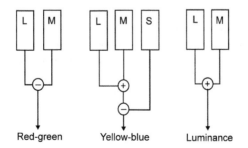

31. **Cone contributions to colour opponent ganglion cell responses**

clusters of neurons known prosaically as 'globs'. A feature of responses from these 'higher' areas is that they are largely indifferent to the colour of the light illuminating the scene: this is referred to as colour constancy. A scene at sunset, or a room lit by tungsten light, has a quite different overall wavelength distribution than one lit by noonday sunlight, and yet the colours appear almost unchanged. It seems that this is achieved by comparing the hue at a particular point with the colour averaged over an area surrounding it. Since this larger area takes in features of many hues it can be taken as approximating to white, and the local hue can be appropriately modified by input from this 'white' surround. Most modern cameras incorporate a system that corrects for the colour of the light illuminating the scene in a similar way; older film cameras did not, and required a different film for 'tungsten' and for daylight. Only under monochromatic light, from a sodium street lamp for example, does colour constancy break down.

Colour science

Lights and pigments behave differently when mixed. When lights of two different wavelengths are added, the result is simply the sum of the two spectral distributions, and is treated by the cones accordingly. Thus a blue light (440 nm) and a yellow light (580 nm) give a spectrum which stimulates all three cone types (Figure 30), and the result is white. Mixing yellow and blue pigments, however, gives a green colour. The reason for the difference is that pigments absorb what they don't reflect. So a yellow pigment reflects some green and red but no blue, and a blue pigment reflects blue and some green but no yellow. What you then see when the pigments are mixed is what is left in the combined spectrum of the reflected light, which is only green. Thus the colours of lights are additive, but those of pigments are subtractive.

Describing a colour is not simple, and over the past century many different methods have been devised to quantify the colours we see. They are all based on a three-dimensional 'colour space' such as that shown in Figure 32. The vertical axis is the overall brightness level, from dark to light. The central circular plane has two dimensions: the circumference represents the pure hues of the spectrum, with red joining up with blue to give the non-spectral colour purple. The central point where the red-green and yellow-blue diameters intersect is white, because here there are equal amounts of the primary colours. Going outwards from the centre towards the circumference the hue components become stronger, and the white component weaker: the colours become more saturated. Thus a pale pink or a pale blue are represented quite close to the centre, on radii joining the centre to the red and blue hues respectively. The whole diagram is really a sphere, since each colour plane is reproduced at every brightness level: when all is dark or overwhelmingly bright no colours can be seen, so the colour plane shrinks to zero at the top and bottom of the brightness axis.

The diagram in Figure 32 is illustrative but not quantitative. In 1931 the Commission Internationale de l'Éclairage devised a scheme, known as the CIE Chromaticity Diagram, which did provide a quantitative system in which any colour could be represented by a pair of coordinates. The axes of this two-dimensional plot are ratios derived from the quantities of three primary illuminants, and it is this use of ratios that allows what is really a three-dimensional space to be collapsed into two dimensions. This system has universal validity, but is not straightforward to use, especially for such mixtures of colour pigments as are used commercially. In 1905 Albert Munsell devised a system with three coordinates that could be reproduced as surface colours, and in 1929 produced *The Munsell Book of Color*, which is still widely used, not just for paints but for the colours of skin, hair, soil, teeth, and beer. The coordinates Munsell used to specify a colour were Value, Hue, and Chroma, corresponding approximately to the Brightness, Hue, and Saturation of Figure 32.

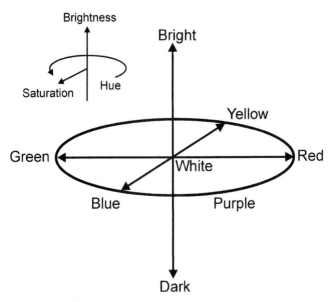

32. Perceptual colour space

Other animals

As mentioned earlier, the mammalian colour vision system has fewer cone types than many other vertebrates. Birds typically have four cone pigments, often including one whose spectral sensitivity extends into the ultra-violet. They also make use of coloured oil droplets in the cones. These droplets act as filters in the light path that narrow the spectra the cones receive, and so improve their wavelength resolution. It is likely that birds, reptiles, and bony fish have finer colour discrimination capabilities than humans.

In 1914 Karl von Frisch showed that bees could be trained to perceive different colours, and then in the 1920s Alfred Kühn found that they could also be trained to recognize ultraviolet wavelengths. We now know that bees, like us, are trichromats, but with a spectrum shifted about 100 nm towards the shorter wavelengths: their three visual pigments absorb maximally at 350 nm (ultraviolet), 440 nm (blue), and 540 nm (green). Trichromacy is common in insects and this has been exploited by flowering plants for the mutual benefit of both—flowers supply nectar and insects pollinate the flowers. Ultraviolet photographs of flowers often show markings on the flowers—nectar-guides—which are invisible to us but make it easier for bees to find a flower's centre. Some butterflies, which use colour for display as well as to find food, can have as many as five visual pigments.

The record for the greatest number of visual pigments is held by the mantis shrimps (Stomatopoda), an ancient order of highly predatory crustaceans that mainly inhabit coral reefs. Their remarkable eyes contain up to 12 visual pigments in a band across the centre of the eye, which is devoted to colour and also to polarization vision. Eight of these pigments cover the visible spectrum, and, as in birds, these are supplemented by filters which narrow their spectra; the other four pigments span the ultraviolet range from 300 to 400 nm. As it is fairly inconceivable that these animals have a 12-dimensional colour space, they must determine colours by doing something other than taking ratios as humans and bees do. Perhaps they can simply say 'it's number 5 so it's yellow'.

Certain snakes are capable of seeing in the infrared, the long-wavelength part of the spectrum beyond 800 nm. Infrared radiation is warm, and is used by thermal imaging cameras to detect humans in the dark. Snakes use it for the same purpose, to detect rodent prey. They do not use the eyes for this, but special pits with heat sensitive nerve endings. These pits are effectively pin-hole cameras and although their resolution is poor they allow the snakes to home in on anything warm. Mammals can't use this trick because of the veiling heat from their own warm bodies.

Chapter 7
Seeing and the brain

The visual pathway

One might think that the optic nerve would go directly to the cortex, where a neural version of the image in the eye would be recreated, turned the right way up, and that would be what we see. It turns out not to be that simple. When it leaves the bony orbit, each optic nerve splits into two halves, each corresponding to the right or left half of the visual field (Figure 33). The two left halves (shaded), each representing the right half of the visual field because of the inversion of the image in the eye, now join together at the optic chiasm, and proceed to the left lateral geniculate nucleus (LGN). Similarly the two right halves representing the left side of the visual field go to the right LGN. The LGNs are structures in the thalamus, deep in the brain, and they relay the signals from the retinal ganglion cells, without a great deal of change, to the primary visual area on the inside of the occipital lobe at the back of the cerebral cortex. This area is also known as the striate cortex, from its layered structure, or simply as V_1.

The reason for the complicated splitting and crossing of the optic nerves is to bring the corresponding parts of the image from the two eyes together, so that they can be compared in the cortex, and disparities between them detected to produce depth (see Chapter 6). In non-mammalian vertebrates the optic nerves do

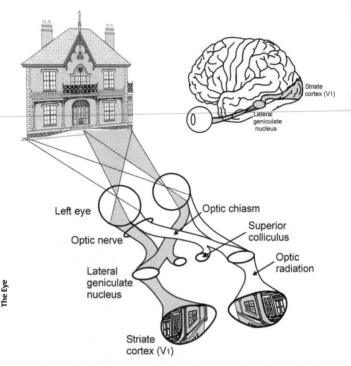

33. The early visual pathway. Each optic nerve splits after leaving the eye, with the two left and two right halves combining, before proceeding to the left and right striate cortex via the lateral geniculate nuclei. The cortical representations are distorted, with the foveal regions greatly enlarged compared with the retinal periphery. The inset shows the pathway from the side. The lateral geniculate nuclei are part of the thalamus, inside and below the cortex

not split but simply cross, with the left eye projecting to the right brain, and vice versa.

Part of each optic nerve goes to the superior colliculus, rather than the cortex. This mid-brain structure pre-dates the cortex in vertebrate evolution; in non-mammals it is known as the optic tectum, and it was the main brain region involved in vision (see Chapter 5). The tectum was where visual information guided motor behaviour: swimming, walking, and flight in birds. The role of the superior colliculus in mammals is mainly related to eye movements and the visual stabilization of gaze; the parietal lobe of the cortex (Figure 37) has largely taken over its function in the coordination of action.

The striate cortex—area V1

It took a long time to work out what was happening in the visual cortex, basically because researchers were looking for concentric receptive fields that were like those of the retinal ganglion cells (Chapter 3). The breakthrough came in 1958 when David Hubel and Torsten Wiesel discovered, more or less by accident, that the cells of V1 were not interested in concentric stimuli, but overwhelmingly preferred lines and edges. This passage from Hubel's book *Eye, Brain, and Vision* provides a wonderful example of how scientific discoveries really come about.

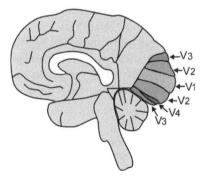

34. View of the *medial* face of the right cerebral hemisphere, showing the location of the secondary visual areas that surround V_1. They are mostly deep within the cortical folds, and not visible from the brain surface, but do protrude on the medial surface. Area V_5 remains out of sight

After about five hours of struggle, we suddenly had the impression that the glass with the dot was occasionally producing a response, but the response had little to do with the dot. Eventually we caught on: it was the sharp but faint shadow cast by the edge of the glass as we slid it into the slot that was doing the trick. We soon convinced ourselves that the edge worked only when its shadow was swept across one small part of the retina and that the sweeping had to be done with the edge in one particular orientation. Most amazing was the contrast between the machine-gun discharge when the orientation of the stimulus was just right and the utter lack of a response if we changed the orientation or simply shined a bright flashlight into the cat's eyes.

The cells in V_1 are primarily concerned with oriented contours—lines and edges—and are organized in columns at right angles to the cortical surface (Figure 35). The input from the LGN is still in the form of unoriented concentric receptive fields like those of the retinal ganglion cells: on-centre off-surround, or vice versa. However, above and below layer IV, where the input arrives, these neurons with circular fields become grouped into aligned assemblies, so that, for example, a row of on-centre cells converges onto a neuron that becomes a line detector with a preferred 'on' region flanked by 'off' regions (Figure 35a). Each column contains neurons with a preference for a contour with a particular orientation in the image; the adjacent column will prefer an orientation a few degrees clockwise or anticlockwise, so that across about 1 mm of cortex all orientations are represented (Figure 35b).

There are two other types of representation in V_1. The inputs from the two eyes are still separate, so that cutting across the array of orientation columns are rather larger columns dominated by one eye or the other. It is not clear at present whether differences between the images—disparities that give rise to depth—are extracted in V_1 or elsewhere in the cortex. In addition, colour information from the colour-opponent cells of

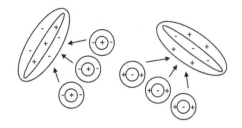

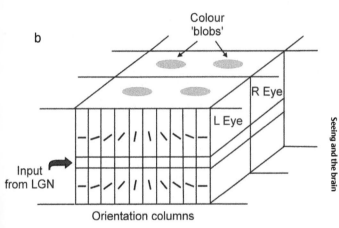

35. Organization of the visual cortex. (a) Oriented line and edge detecting cortical cells are derived by the addition of the outputs of LGN neurons with circular centre-and-surround receptive fields. (b) This 'ice-cube' model of a hypercolumn represents a single pixel in the cortical image. In fact the column structure is more fluid: wheel-like rather than linear

the retina is present, but largely kept separate from contour information. This information is sequestered in 'blobs'—cylindrical structures that penetrate the array of columns. Again it is unclear how much processing of colour occurs in V₁ and how much in other areas such as V₄. The whole assembly of orientation columns, eye dominance columns, and colour blobs is known as a hypercolumn, and represents one pixel in the analysis of the image (Figure 35b).

The impression one has of the 'image' that V1 extracts is that of a cartoon, made up of short monochrome lines and edges. Although devoid of tone, colour, and depth, such an image is quite intelligible and defines most of the important features of a scene, even of a face (Figure 36). The image in Figure 36b was made by Don Pearson and his colleagues in the 1980s, originally as a way of transmitting images for the deaf over the then extremely limited channel capacity of telephone lines. They found that the best detectors for representing the main features of real faces were 'luminance valley detectors', each consisting of a 5 x 5 matrix of photo-detectors in several orientations (Figure 36c and d). These find dark line segments on a light background. The output of the detectors was then reproduced as black dots on a screen, and this could provide moving images at up to six frames a second. This is enough to convey the identity of the sitter, and their facial expressions and mood. Pearson's detectors can be thought of as executing operations similar to the off-centre line detectors in V1 (Figure 35a). The cartoon image in Figure 36b gives a feeling for the kind of representation in V1.

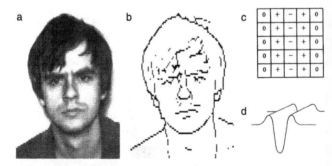

36. 'Cartoon' reconstruction of a face (a and b) made by passing the original photograph through a matrix of 'luminance valley detectors' (c and d), analogous to the off-centre line detectors in V1. See Figure 35a, right

Surrounding V₁ and anterior to it are other cortical areas that get their input from V₁, and perform further analyses on the image (Figure 34). The anatomist David Van Essen estimates that there are 10–12 cortical areas that, like V₁, have an organization which corresponds in a map-like way with the retinal image. Thus there appears to be no single straightforward complete representation of the visual world in the cortex, but a series of modules each with a partial but distinct role in producing the image we perceive. A problem is that the evidence for this comes from two different sources: micro-electrode studies of the monkey cortex, and functional magnetic resonance imaging (fMRI) of human subjects. The correspondence between these two sets of results is surprisingly good, but not good enough to make exact matches between specific areas in the two species. Nevertheless there are about four areas, in addition to V₁, that most agree are distinct. Area V₂ has a role in analysing disparity and in distinguishing simple shapes; V₃ responds to coherent motion of large parts of the visual field; V₄ has a more complex role than V₁ in the analysis of shape and colour; and V₅ is concerned with analysing the speed and direction of moving objects. Damage to V₅ can result in the very rare condition of *akinetopsia*, or motion blindness, in which movement is seen not as motion but as a series of static images. This makes it difficult to pour tea into a cup, or cross a road safely. Like *achromatopsia* (Chapter 5) this demonstrates that damage to a single cortical module can affect a specific aspect of the overall percept. Some authors distinguish many more visual modules in the cortex. The situation is a little like botanists in the 19th century who were either lumpers or splitters, amalgamating genera or creating further divisions.

Dorsal and ventral streams

In 1982 Leslie Ungeleider and Mortimer Mishkin proposed that the flow of information out from V₁ took two broad routes: one in the direction of the temporal lobe and the other towards the parietal lobe (Figure 37). They referred to these as the ventral

'What' and dorsal 'Where' pathways. The 'What' pathway was concerned with the identity of objects and the 'Where' pathway with object location. In the 1990s David Milner and Mel Goodale modified these ideas so that the ventral pathway became identified with perception, and the dorsal pathway with the control of action. Much of the evidence for this division came from patients who had deficits in either perception or action. For example, a patient D.F., who had a lesion in part of the ventral stream, was asked to match the orientation of a card she held to the angle of a slanted slot in a board; she was quite unable to do this. However, when asked to post the card into the slot, she had no difficulty doing so. Her problem was thus one of perception (*visual agnosia*) but not motor control. Conversely there were patients who had no difficulty with the perceptual matching task, but were unable to perform the posting task (*optic ataxia*). They all had lesions in the parietal region of the dorsal stream. It is interesting that D.F. was also unable to give a verbal report about the orientation of the card, whether or not she was posting it. On the basis of this and other studies Milner and Goodale concluded that conscious awareness of visual phenomena requires an intact

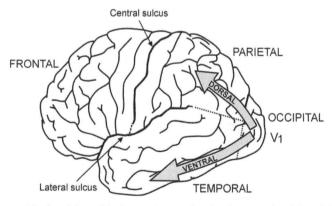

37. The four lobes of the human cortex, showing the ventral and dorsal information streams originating from the occipital lobe at the rear of the cortex

ventral stream, and that activity in the dorsal stream is largely unavailable for conscious report.

Face recognition

It had been known since the 1960s that there are cells in the temporal lobes of monkeys that are uniquely responsive to faces. Cells may respond to a face independent of the lighting conditions, to a face in profile but not frontally, or to the face of a particular person. There seems to be a complex hierarchy of cells dealing with different aspects of face recognition. Related to this there is a condition that affects some people, described by the neurologist Oliver Sacks in his book *The Man Who Mistook his Wife for a Hat*, which renders patients incapable of recognizing faces. This condition—*prosopagnosia*—is the result of temporal lobe damage, often caused by a stroke. Faces are geometrically very complex, and we can recognize many thousands by very subtle and barely describable differences. It is not known how this comes about, although the monkey studies suggest that some problems, such as the trade-off between size and distance, and the changing effects of light and shade, have already been dealt with on the way to the temporal face-processing areas. More worrying is the question of where other recognizable objects are represented. Despite many attempts to identify cells in monkeys that respond uniquely to particular objects, other than faces and possibly hands, none have been found. There are regions in the ventral stream that respond to geometrical configurations such as stars, ovals, and more complex feature clusters, but these are not structures with an obvious meaning in the world. It is understandable that sociable animals such as primates should prioritize faces, and perhaps have special mechanisms for dealing with them, but what about leaves, mangoes, and leopards? Perhaps the right questions have yet to be asked.

I cannot resist mentioning that sheep also recognize faces—of sheep. In a series of studies Keith Kendrick and his colleagues showed that sheep can learn to recognize up to 50 individual sheep faces. They can also recognize humans, but not as individuals. Like primates, sheep have a 'face area' in the temporal lobe. Many cells there respond to any face with horns, others respond only to faces of familiar Dales bred sheep, and a much smaller number respond to both humans and dogs, without distinguishing between them. Presumably they only need one category for the 'bossy non-sheep animal'!

Other brain regions

About 27 per cent of the human cortex is predominantly visual in function. This compares with 8 per cent for audition, 7 per cent for the body surface, and 7 per cent for motor action. In macaque monkeys the visual proportion is even higher, 52 per cent, but this difference is mainly due to the much larger frontal lobes of humans, associated with cognition, emotion, and language. In addition to the specifically visual areas there are many others that are involved less exclusively in visual processing. Eye movements, which are crucial to the way we take in information, are organized at the cortical level by the frontal eye fields towards the rear of the frontal lobe, in cooperation with a specific region in the parietal lobe (Chapter 4). In the parietal lobe of monkeys there are defined regions devoted to reaching and to grasping; these pass dorsal stream visual information on to the pre-motor cortex in the frontal lobe and then to the motor cortex just posterior to the central sulcus (Figure 37), from which commands to the muscles are initiated. No doubt more complex visually demanding activities such as ball sports or driving have a similar frontal-parietal organizational basis.

Social interaction is another aspect of primate life that requires a high visual input. Indeed the large size of the frontal lobes in primates and particularly man is often attributed to the complexity of living cooperatively in large groups. The face areas of the temporal lobe are obviously important here, but so are more basal structures such as the amygdala, a nucleus deep inside the temporal lobe. This structure is activated by images of fearful and angry faces, as well as emotions associated with aggression and sexuality.

Another important aspect of behaviour that relies on visual information is the ability to locate oneself and to navigate through the world. In the 1970s John O'Keefe found that cells in the hippocampus, another structure deep in the temporal lobe, responded only when a rat was in a particular place in a maze. These 'place cells' appear to form a map system which animals including humans use as a basis for navigation. Supporting this, in 2000 Eleanor Maguire and her colleagues found, using magnetic resonance imagery, that London taxi drivers, who spend two years or more acquiring 'The Knowledge' of London streets, had a small but significant expansion of the posterior hippocampus compared to control subjects. In a similar vein, food-hoarding birds such as jays, which have to remember what has been stored where, have a larger hippocampus than non-hoarders. A map by itself does not tell you whether to turn left or right, and as every map reader knows it has to be translated into 'egocentric' coordinates, to provide directions relative to the body, before it can be used to navigate. It seems that this egocentric coordinate system is located in the parietal lobe, where it can feed into the motor system. It is also from the parietal lobe that visual information from the dorsal visual stream can be fed back into the map system of the hippocampus.

Using visual information: the role of attention

There is far too much information in a visual image to be analysed fully during a single fixation, and in any case most of it is not important for a current course of action. Whenever vision is deployed in the execution of a task there need to be arrangements for highlighting those features of the scene that are relevant, and downplaying those that are not. These are the attention mechanisms, and they operate 'downwards' from the parts of the cortex where tasks are organized—primarily the frontal and parietal lobes—to the earlier regions of the dorsal and ventral streams that process the visual information. There are many cells in the visual regions of the cortex that increase their firing rate when a monkey is engaged in a particularly visually demanding task, so a descending attention system is certainly present. There is, however, no clear information about where the neurons that do this tweaking originate, although there is no shortage of possible anatomical pathways. The more serious problem is one of addressing the right targets. Consider a carpenter about to use a hammer and a nail. The possible locations of the required objects need to be highlighted in spatial memory—wherever that is kept—so that the carpenter can go straight to them, rather than having to search. The clusters of features that will identify the hammer and nail also need to be singled out from the feature catalogues of the ventral stream. It seems that for each task a rather complex attentional template has to be in place, and each template will be different depending on the combinations of object and action involved. There has been much progress in the last half century in determining how the rear half of the cortex analyses incoming visual information, there remains a huge task in working out how this information is usefully deployed.

Development

Our visual system is not just an unfolding product of our genes: as it develops it is also moulded by the visual information it receives. In his Nobel Lecture in 1981, the Swedish neurophysiologist Torsten Wiesel summarized early visual development:

> Innate mechanisms endow the visual system with highly specific connections, but visual experience in early life is necessary for their maintenance and full development.... Such sensitivity of the nervous system to the effects of experience may represent the fundamental mechanism by which the organism adapts to its environment during the period of growth and development.

During this developmental period the competitive processes that occur between cortical neurons can lead to a well tuned visual system, but also to long lasting problems. If one eye is deprived of vision in infancy, by a cataract or some other occlusion, the other will take over V_1, becoming dominant, and it will remain so throughout life. If the eyes are not aligned, as in a squint, binocular connections do not develop, and depth perception using disparity is precluded. Treatment of these conditions has to be carried out as early as possible, while cortical connections are still being refined.

There are many stages to human visual development. A newborn baby will orient to simple targets, largely using the ancient sub-cortical visual system. It takes three months before the cortex is sufficiently well developed for object recognition, and for the proper control of eye and head movements. Reaching for seen objects comes after six months, and visual control of locomotion at about a year, when an infant begins to explore space beyond its immediate reach. Speech begins about six months later. At three weeks after birth visual acuity is less than a 20th of its adult value, after ten months this has risen to about one-third. It continues to improve until the age of about 3, due to changes in both the retina and the cortex. By about 18 months the perceptual and attentional machinery of the brain is already well developed. The dorsal stream seems to develop more slowly than the ventral: catching a beach ball takes several more years to perfect. In an old-time apprenticeship, learning the visuo-motor skills of a trade took seven years, and sportsmen can take even longer to fully master their game. Visual learning of one sort or another continues throughout life, but the foundations on which the rich edifice of adult vision will be built are present from the age of about 2.

Chapter 8
When vision fails

Loss of sight is something that most of us dread, as it means the loss of our ability to move around freely and deal with the objects of everyday life. Failing eyesight takes many forms, from age-related conditions that reduce acuity or field of view, to complete blindness resulting from accident or disease, and in the worst case to blindness from birth caused by mishaps in development. The eye is not the easiest organ to repair when problems arise, but much can be done to improve the eye's optics or to stop retinal deterioration from getting worse. Even when blindness has set in, as a result of currently incurable conditions such as macular degeneration or retinitis pigmentosa, there are ways of stimulating the visual system via electrically powered implants, which can provide a useful degree of vision. Other approaches, for example the use of stem cell therapy, are beginning to show promise in animal trials. In this final chapter, I will first review the major conditions that may lead to degradation or loss of vision, and then outline some of the newer approaches to providing some useable visual input to those whose eyesight has been lost.

Conditions that threaten eyesight

In early life most visual problems are optical, for example short and long sight, as discussed in Chapter 3. About 15 per cent of children under the age of 15 have these focus problems. All are easily dealt with using glasses or contact lenses. Between the ages of 40 and 50 presbyopia, in which the lens becomes too stiff to accommodate to different distances, becomes a near-universal condition. Once again, glasses are the simplest answer. Later in life cataracts, in which the lens starts to crystallize and become opaque, begin to degrade vision. By age 75 about 30 per cent of people have cataracts. The remedy is an operation in which the lens is removed and replaced with a plastic one. This is a very common procedure, and nearly always successful.

Several diseases can affect vision in later life. Diabetes is one of the main causes of visual problems, and if untreated for 15 years it can lead to impaired vision in 10 per cent of cases and blindness in 2 per cent. Both Type I (insulin dependent) and Type II (late onset) diabetes result in elevated levels of blood glucose, and this in turn causes the retinal blood vessels to leak and ultimately burst. They may then start to proliferate, making things worse. The leaks cause blurring and opacities that block vision, and may also lead to retinal detachment. Treating the disease itself to reduce blood glucose levels, by the use of insulin or other drugs, is the best preventative measure. However, if the disease has taken hold the usual treatment is to destroy proliferating blood vessels using laser surgery. This is done in a grid pattern in the peripheral retina in order to keep the central retina free of blood and debris. Laser surgery is not a cure, but it can be very successful in preventing further damage.

Glaucoma is a group of diseases that is usually caused by abnormally high pressure inside the eyeball. It affects about 2 per cent of over-50s, rising to 10 per cent in over-80s. The elevated pressure has the effect of squeezing the optic nerve, damaging the fibres and restricting blood flow to the optic nerve head. The outer fibres are affected first, and as these tend to come from the more peripheral parts of the retina sight is first lost from the outer parts of the visual field (Figure 38). This loss gradually spreads towards

38. Appearance of the image in glaucoma and age-related macular degeneration

the centre, and can eventually lead to irreversible blindness. If detected early, glaucoma can be treated easily with eye drops that reduce the intra-ocular pressure. The early detection of both glaucoma and macular degeneration (see later in the chapter) has been greatly assisted in recent years by the technique of optical coherence tomography, which provides a 3-D image of the intact retina, and shows up both the normal layering, and any pathological abnormalities.

Two more sinister diseases are age-related macular degeneration (AMD) and retinitis pigmentosa (RP). In AMD damage is caused by the accumulation of plaque material, known as drusen, in the region of the macula in the centre of the retina. This build-up, behind the retina, can destroy the pigment epithelium and photoreceptors, and also cause retinal detachment. Ten per cent of sufferers have the more serious 'wet' form of the disease, in which blood vessels proliferate behind the retina, causing leakage and further damage to the photoreceptors; some drugs are now available that can curtail this proliferation. The macula is the region around the fovea, and is crucial for reading, face recognition, and other fine discrimination tasks, so its loss is devastating, even though peripheral vision is usually unaffected (Figure 38). The problem is that half the primary visual cortex is devoted to the macular region, and so it is very hard to use peripheral vision for tasks usually performed with central vision. Acuity is much lower outside the macula, and even with substantial magnification, reading is very difficult. AMD is not uncommon: 10 per cent of people between 66 and 74 will have it to some degree, and this rises to 30 per cent above the age of 75. AMD is very variable, so the degree of impairment may be minimal in some cases but debilitating in others. Although there are no effective treatments for AMD, dietary supplements can sometimes help. Recent trials in mice using replacement photoreceptors derived from human embryonic stem cells have shown promising results.

Retinitis pigmentosa is the name for a group of inherited diseases that lead to a progressive loss of photoreceptor cells, and hence visual performance. RP is fortunately rare, affecting fewer than 1 in 3,000 of the population. Unlike the conditions discussed above, RP begins to show itself before the age of 30, and becomes progressively worse. In 1969 a mutation was found in the gene that codes for rhodopsin, and since then about 150 mutations of this gene have been linked to RP. The rods degenerate first, and this leads to poor night vision and loss of peripheral vision. Loss of day vision, involving the cones, only becomes apparent in the last phases of the disease, and this may occur by the age of 50. Large doses of Vitamin A (the precursor of the chromophore retinal; see Chapter 1) are said to be helpful. Although there is currently no accepted treatment for RP there are promising developments. Gene therapy, in which the defective gene is replaced with a normal one using a viral vector, has been the subject of clinical trials since 2007. Stem cell research may also lead to a replacement for photoreceptors at some time in the future. However, because it often leads to blindness, other therapies such as the electrode-base prostheses discussed later, become realistic treatments.

Electronic prostheses

In recent years devices have become available which can supply electric signals directly to either the retina or the brain. These signals typically result in the appearance of spots of light, and are thus not in the form the brain is accustomed to dealing with, which means that the normal machinery for dealing with images is only partially available. For example, tasks such as resolving objects against a cluttered background are much more difficult when the brain's input is in the form of an array of light spots, rather than a complete image. When the retinal receptors no longer function, as, for example, in retinitis pigmentosa, but the rest of the visual nervous system is intact, it may be worthwhile implanting an electrode array to provide a stimulus pattern that bears at least some resemblance to the retinal image. Three locations for such an array have been tried: the retina itself, the optic nerve, and the primary visual cortex.

Retinal implants. Probably the most successful implants, and the only ones currently licensed for general medical use, are epi-retinal prostheses. These are arrays of electrodes placed on the outer—ganglion cell—face of the retina. An advantage of such a device is that it uses the retina's own output stage to supply information to the brain, and so preserves the geometry of the ganglion cell layout. The information provided by the array comes from an external camera. The first 16-electrode array was implanted in the eye of a patient at Johns Hopkins University in the early 1990s. Six more patients had the implant in the early 2000s, and it is claimed that from being completely blind they could now perform a range of vision-based tasks. A second generation of implants, using 60-electrode arrays and known as Argus II, was trialled in 2007 and is now approved for use in the US and Europe.

Sub-retinal prostheses are another active field of research (Figure 39). These are located behind the retina, in front of the pigment epithelium, in the space occupied by the now-defunct receptors. A team at the Eye Hospital of the University of Tübingen produced an array of photodiodes, boosted by an external power supply, which supplied electrodes that stimulate the retinal bipolar and ganglion cells. The current generation of

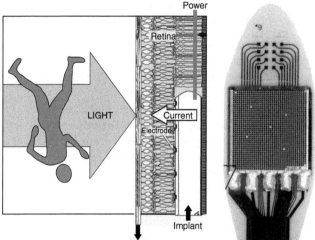

39. Sub-retinal implant. Left: the implant in place in the space previously occupied by the receptors, between the pigment epithelium and the bipolar cells. Right: the surface of the implant, with a matrix of 1,500 electrodes

implants uses a chip with 1,500 electrodes. The photodiodes make use of the image in the eye itself, rather than an external camera, and so preserve the geometry of the visual image. Since the array moves with the eye the wearer can select where to look, as in a normal vision. In a clinical study of 11 blind patients, some were able to read letters, recognize unknown objects, and localize a plate, a cup, and cutlery. All patients had some degree of sight restored. This system shows great promise, but at the time of writing it is not in general use.

Optic nerve implants. A prosthesis under development in the University of Louvain uses an array of electrodes on a cuff around the optic nerve, fed by signals from an external camera. This again produces light phenomena seen by the patient, but these are not particularly easy to interpret because of the lack of a clear relationship between the positions of the electrodes and the geometry of the retina.

Cortical implants. As early as 1968, Brindley and Lewin published a study of a 52-year-old patient who had recently become blind. She was provided with an implanted array of 81 electrodes in contact with the surface of the primary visual cortex. The array was supplied from an external camera and activated via radio transmitters. Surface electrodes were used rather than ones that penetrated the tissue to avoid problems of chronic inflammation, but they needed large currents—milliamps rather than microamps—to produce visual effects. According to the patient, activating the electrodes produced spots of light 'the size of a grain of sago at arm's length'. Only about half the electrodes produced these 'phosphenes', and the use of the array for recognizing patterns was very limited. Since this early study, others, notably at the University of Utah, have developed arrays of finer electrodes that penetrate the cortex and use far less current. The electrodes are made of silicon with platinum-coated tips, materials which are unlikely to cause inflammation. So far these are still under development, and there remain questions of long-term safety and durability. It is difficult to predict how long it will take for cortical prostheses to become useful, but making devices that will allow reading and a degree of mobility are the goals, and the prospects seem good in the medium term. For the time being, however, retinal implants are the only devices that can be routinely used.

Patients who are blind from birth

The devices discussed in the last section are really only available to patients who have gone blind in the course of their lifetime. They have, essentially, a nervous system that is still capable of making sense of visual input. For those blind from birth this is unlikely to be the case. We know from half a century of physiological experiment that the brain requires vision in infancy to become properly wired up. Richard Gregory, in his book *Eye and Brain*, reports the case of a man who had been blind from the age of 10 months due to corneal opacity, and had sight restored at the age of 52. He had enormous difficulty in matching up his new visual impressions with the tactile world with which he was familiar. He never learnt to read by sight, although he could read both Braille and block letters by touch. He preferred to sit in darkness, became depressed, and died three years later. For those blind from birth, sight restoration is not an option.

Most attempts to produce aids for the profoundly blind concentrate on substituting some other sense for vision. White sticks have long been used to produce clicks whose acoustic echoes contain limited distance information. In the 1960s Paul Bach-y-Rita developed a system in which a video image was supplied to an array of activators that created an equivalent tactile image on some part of the body. This could be the back, chest, fingertip, abdomen, or forehead (Figure 40). The system is very successful, and after 50 trials subjects could learn to recognize simple shapes in various orientations with 100 per cent accuracy. Various systems have been devised that convert a video image into a 'soundscape'. For example, these can convert height to pitch and brightness to loudness and can scan the scene using a head-mounted camera. Other devices provide distance information via an echolocation system based on clicks to the two ears—a technological advance on the white stick.

There is evidence from fMRI scans that in blind people the visual cortex is taken over to some extent by other senses. So, for example, the processing of objects perceived through the sense of touch takes place, in part at least, in what for sighted people would be the visual cortex; thus in some sense these objects are 'seen'. This capacity for sensory substitution in the brain seems to offer much scope for providing the blind with alternative sources of imagery.

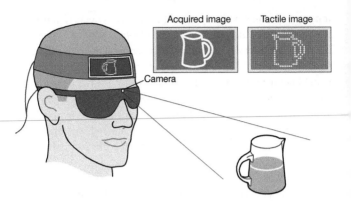

Acquired image   Tactile image

Camera

40. Sensory substitution. The image from a camera worn on eye-glasses is converted to a tactile image by mechanical activators on a band round the forehead

Computer vision

As a postscript to this discussion, I will briefly mention visual systems that are entirely synthetic, and do not require a human brain to process an image. Humans are able to guide themselves through the environment, recognize objects and people, and manipulate tools and materials, all with apparent ease. This led early computer engineers, in the 1950s and 1960s, to think that duplicating such performances using cameras and computational power would be a straightforward task. It has proved not to be so, but knowledge acquired in the half century since the enterprise began has finally produced devices such as self-driving vehicles on the roads of California, and face-recognition security systems in use in airports around the world.

With hindsight it is easy to see why it has taken such a long time for artificial vision to come of age. Consider face recognition. In airport security systems you are photographed looking directly at the camera, and lit by diffuse light. Such an ideally presented face can then be compared with a stored catalogue of similarly obtained pictures, and a match found by statistically comparing geometrical features. This is not too hard a task, and one that computers are good at. However in real life faces come in all orientations from full-face to profile, their size depends on distance, and the direction and nature of the illumination completely changes the pattern of light and dark presented by the features. They may, in addition, have expressions such as smiles or frowns that distort the 'standard' pattern of the features. They may have to be extracted from a cluttered background or a crowd of other faces. The number of variables is now far too high for current computational systems to deal with, and 'spotting a villain in the crowd' software is hardly ever successful for this reason. Humans are, however, remarkably good at recognizing faces despite all these complications, but *how* we do it still remains a major unsolved problem. There is no guarantee that when computer engineers do come up with answers to some of these problems their solutions will bear much resemblance to the methods the brain uses, but at least the questions will be the same.

Further reading

Chapter 1: The first eyes

M. F. Land and D.-E. Nilsson (2012) *Animal Eyes* (2nd edition). Oxford University Press.

P. Holland (2011) *The Animal Kingdom: A Very Short Introduction*. Oxford University Press.

Chapter 2: Making better eyes

M. F. Land and D.-E. Nilsson (2012) *Animal Eyes* (2nd edition). Oxford University Press.

Chapter 3: The human eye

C. W. Oyster (1999) *The Human Eye*. Sinauer Associates.

Chapter 4: The moving eye

J. M. Findlay and I. D. Gilchrist (2003) *Active Vision*. Oxford University Press.

M. F. Land and B. W. Tatler (2009) *Looking and Acting*. Oxford University Press.

Chapter 5: The third dimension

G. Mather (2009) *Foundations of Sensation and Perception* (2nd edition). Psychology Press.

Chapter 6: Colour

A. Valberg (2005) *Light Vision Color*. John Wiley & Sons.

J. D. Mollon, J. Pokorny, and K. Knoblauch (eds) (2003) *Normal and Defective Colour Vision*. Oxford University Press.

O. Sacks and R. Wasserman (1987) *The Case of the Colorblind Painter*. New York Review of Books, Vol 34 No 18.

Chapter 7: Seeing and the brain

A. D. Milner and M. A. Goodale (2006) *The Visual Brain in Action* (2nd edition). Oxford University Press.

Chapter 8: When vision fails

World Health Organization website on blindness and visual impairment: <www.who.int/blindness/en/>.

Royal National Institute for the Blind website: <www.rnib.org.uk>.

The Eye

Index

Index

Expand your collection of
VERY SHORT INTRODUCTIONS